Masonry and Plastering

Other titles in the Crowood DIY series

MASONRY
& PLASTERING

Mike Lawrence

The Crowood Press

First published in 1991 by
The Crowood Press Ltd
Ramsbury, Marlborough,
Wiltshire SN8 2HR

British Library Cataloguing in Publication Data

Lawrence, Mike *1947–*
    Masonry and plastering.
    1. Buildings. Walls
    I. Title
    693

ISBN 1 85223 517 9

Acknowledgements

Line-drawings by Andrew Green.

The author would like to thank the following companies and organizations for
providing the photographs listed below:

Bradstone Garden Products (pages 8, 43, 44, 51 and 85);
Robert Harding Picture Library (pages 10, 11, 12, 13, 14 and 15 (top));
Marshall Cavendish (pages 7, 15 (bottom), 16, 17, 18, 31, 41, 53, 57, 59, 61, 63, 68, 69,
87, 90 and 92);
Marshalls Mono Ltd (pages 35 and 39).

Typeset by Acūté, Stroud, Glos
Printed and bound by Times Publishing Group, Singapore

Contents

Introduction

The phrase 'bricks and mortar' is a modern euphemism for house and home; it is what every home owner spends a working life investing in, and it represents the biggest asset most of us ever own. Few people actually spend their lives laying bricks and mortar, but there are plenty of opportunities for the amateur to dabble in bricklaying and masonry work – especially in the garden – and in related crafts such as plastering. Indeed, with the high cost of employing professional craftsmen of all types, there is every incentive for the do-it-yourselfer to master these skills and to marry them to his or her own creative talents. The end product may be nothing more ambitious than a low garden wall or a feature fireplace in the living room, but it will have the unmistakable stamp of individuality on it.

Most of the building projects covered in this book are outdoor ones, for two reasons. The first is that your early attempts to get mortar to adhere to bricks, and bricks to stack up the way you want them to, are bound to lead to a certain amount of mess, and it is far easier to cope with this in the garden than in the house. The second is that few people actually want to knock their homes about structurally, but long to turn their gardens into a unique feature with a patio to catch the sun, walls and planters to display their favourite plants and shrubs to best effect, and perhaps an ornamental arch leading the way to the vegetable garden. This book will show you some of the things you can easily achieve.

Even if you decide that you would prefer to leave constructional projects to the professionals, you will still be faced with that seemingly endless task of keeping your property – its bricks and mortar, if you like – in sound condition. In the second part of the book you will find advice on everything from repointing walls and patching drives to repairing concrete floors and patching plaster. Each is a small but worthwhile step in maintaining the value of your investment, as well as in making your home a pleasant place to live.

THE BASICS

Most people think of working with bricks and mortar, concrete, paving slabs and walling blocks, even plastering walls and ceilings, as the 'heavy' side of do-it-yourself. Yet there can be few more rewarding ways of spending a day, a weekend, even a season, than in creating a structure of your very own from a formless pile of building materials. As you progress from humble beginnings to more elaborate projects, you will find that there is little mystique in the craft of bricklayer, stone mason or plasterer: all are classic examples of the old adage that practice makes perfect, and practice time will cost you nothing!

This section of the book looks briefly at some of the projects you may want to create, and at the basic skills you will need to carry them out.

What to Tackle Outdoors

If you have never laid a brick or mixed a bucketful of mortar in your life, it is best to start learning by tackling some basic outdoor building projects. Here are some you could consider.

Hard surfaces Whether you have inherited a virgin plot, a neglected wilderness or a reasonably well-kept garden, one job you are sure to want to tackle is the laying (or re-laying) of some hard surfaces to allow you to use the plot in all weathers. These could include a driveway for your car, a path round the house and down the garden to the greenhouse, a patio that catches the sun or allows you to admire the view, even an area for the children to play on when the lawn is too wet. Whichever of these you decide to create, you can use any of a wide range of building materials.

Your choice of materials for each of these surfaces will be decided by a combination of factors. One of the first you will think about is the cost, and somewhat surprisingly you will find that 'seamless' materials such as concrete and macadam cost broadly the same to lay per unit area as paving slabs and blocks (although there are variations in slab and block prices depending on type). The reason lies in the hidden costs of each type – the extras over and above the cost of the material itself.

There are other factors of equal importance, of course. To begin at the beginning, you need to think about how easy the chosen material will be to transport, to move around the site as well as to lay. For example, concrete is hard work to mix and move around in large quantities, while slabs and blocks are easier to handle and can obviously be transported in manageable quantities. Speed of laying, and the time within which the surface can be used, are important too. Concrete takes time to place and needs time to harden, as does macadam, while dry-laid blocks or slabs are quick to lay and can be used immediately. At the other extreme, it can take weeks to point a large area of crazy paving; bedding the pieces is the easy part of the job!

Think too about appearances and future maintenance. Concrete can discolour and break up in time, and major repairs can be difficult to carry out. Macadam can be softened by oil spillages, may become tacky in hot weather and will dent under point loads, but is easy to patch if necessary. Gravel is no good for sloping sites, needs regular raking and weed-killing and can escape onto lawns (or indoors on people's shoes), but looks good and is a wonderful burglar deterrent – no one can tip-toe quietly up a gravel drive! Slabs

Fig 1 (*above*) Acquiring some basic bricklaying skills will allow you to tackle a wide range of creative projects and also to carry out essential home maintenance jobs.

and blocks offer the widest choice in surface appearance and once laid should need little maintenance.

On the subject of looks, remember too that the shapes you want for your hard-surfaced areas may dictate which material you use; seamless materials and small blocks are much easier to lay in curved shapes than large paving slabs.

Apart from hard surfaces to walk, drive or sit on, you may also need areas of hard standing for other purposes – as bases for outbuildings, for example, or for a barbecue in the garden. Here looks are less important than performance, and concrete slabs are often the best choice.

Walls Walls are next on the agenda. What you want to build depends to a large extent on your site; you will obviously have more scope for elaborate construction projects on a sloping site, but even on a level one you can build anything from low-level planters (which also have the advantage of making gardening easier if you have arthritic knees or a bad back) to feature walls dividing up the garden or creating screened and sheltered spots for sitting out. Walls undoubtedly make the best boundaries too, especially at the front of your property where privacy and security are particularly important features, and here the scale of the job gives you plenty of scope for originality.

Your choice of walling materials is even wider than for hard surfaces, ranging from bricks to natural stone (if you can get it or are able to afford it), or the host of imitations of natural walling materials now available. Here your main criterion must be looks; laying a brick, a walling block or a lump of sandstone involves broadly the same technique and takes much the same time – performance is much of a muchness too.

Which one you choose will depend on two things: the sort of garden you want to create, and whether you want to match the materials used to build your house. As far as the first point is concerned, brick walls tend to look much more formal than stone ones, while the second point speaks for itself.

Steps These are a natural progression, especially if you have a sloping site, since they will link individual areas of the garden together once you have used retaining walls to create a series of terraces up or down the slope. Even on comparatively

level sites, you may need steps to provide access to the house. You can set them into the slopes or build free-standing structures; it all depends on the demands of the site. However, in either case steps involve a subtle blend of the building materials you select for both hard surfaces and walls, since their treads and risers combine elements of both.

There are a wealth of other constructional features you can add to your garden, including ornamental planters, archways – free-standing, or incorporated into your dividing or boundary walls – and, perhaps most attractive of all, water features. The sound of running water in a garden is one of the most relaxing things on earth, tinkling merrily away in the sunshine, and even a still pond can be an attractive centre-piece on the patio or in the lawn. It will also allow you to grow a range of aquatic plants, and will attract wildlife of all sorts to your garden – a source of great interest to all, especially to children.

Creating a pond has never been easier, with the range of rigid and flexible pond liners now available to let you form exactly the shape and size of pond you want. Remember too that you are not limited to in-ground pools; there is nothing to stop you incorporating one as a raised feature on your patio or elsewhere in the garden.

Fig 2 (*above*) Your first outdoor building projects are likely to be relatively simple – some shallow steps or an area of paving, for example.

What to Tackle

What to Tackle Indoors

Unless you are a competent enough do-it-yourselfer to tackle major internal structural alterations such as building or demolishing load-bearing walls, you are rather more restricted indoors as far as constructional projects are concerned.

The one major task that you may want to take on is plastering (or replastering) walls and ceilings – a skill that is well worth mastering, because it has such a high labour content. A bag of plaster costs only a few pounds, but employing a skilled plasterer to put it on the wall will set you back quite a sum. It is a skill that can be perfected only with practice, but you will find that after several abortive attempts to make the mix stay where you put it, you will suddenly get the knack! Even if you never get good enough to tackle large areas with confidence, you will still be able to do all sorts of smaller-scale jobs.

You can also use your new-found skill to create arch-shaped openings in doorways and between through rooms, thanks to the availability of wire mesh arch formers. These are rigid assemblies designed to be fitted within the tops of square openings, and come in a range of shapes (see page 66 for more details). Once the former is in place, all you have to do is to apply a skim of plaster over it.

Texturing wall and ceiling surfaces is a skill that is part-plastering, part-decorating. Textured finishes are not only a quick and inexpensive way of covering surfaces in less-than-perfect condition; they give you the opportunity to create your own three-dimensional designs and effects using a material that is both easy to apply and extremely durable in use.

If you want high-relief effects it is best to use a texturing compound (Artex is the best-known, and is a trade name that has become synonymous with textured finishes, as Formica has for plastic laminates). Some come in powder form and are mixed with water prior to being brushed onto the wall, while others are available ready-mixed. For lighter reliefs, you can use a textured emulsion paint instead. This has the advantage of being strippable (using a special chemical stripper) if you want a change of decor in the future; texturing compounds can be removed only with the help of a steam wallpaper stripper. See pages 67–8 for more details.

There is one popular indoor job that will need the bricklaying skills you have acquired from your outdoor building projects, and that is building a feature fireplace surround. With open fires coming back into fashion, many home owners are faced with having to reconstruct fire surrounds that were removed in the 1960s and 1970s (the heyday of the blocker-uppers, when many chimney breasts were reduced to flat-fronted uselessness). One of the most attractive solutions is to use a kit containing special bricks, natural or reconstructed stone, which are available in a wide range of styles and sizes. See pages 60–1 for details.

Maintenance and Repairs

Apart from putting your new skills to creative use, you will also need them for a range of occasional repair jobs around the house. These are perhaps not as much fun, but are just as important to your home's well-being. They are covered in Chapter 4.

Fig 3 (*below*) As your skills improve you can start to build more elaborate projects such as a barbecue.

9

Bricks

Without a doubt the most widely-used building materials are bricks. They have been around for thousands of years, and are as popular today as they ever were. They are versatile, easy to use, and can be very attractive, not least because of the random variations in colour and appearance that most bricks exhibit. Clay bricks are the commonest variety, but you may also come across bricks made from lime and sand (known as calcium silicate, sand-lime or flintlime bricks) or from concrete. There are literally hundreds of different types to choose from, but for practical purposes you can be confident of picking the right brick for the job if you consider three factors: the quality of the brick, its type and its shape.

Quality This is the most important factor, especially since most of the projects you will be working on will be outdoor ones and the bricks you use will have to be weatherproof. Bricks are classified into three qualities to indicate their intended usage. *Internal quality* bricks have no weather resistance, and are suitable only for internal use – rain penetration and frost damage will soon break them up out of doors. *Ordinary quality* bricks can be used out of doors, but are not intended to stand up to severe exposure. You could therefore use them for the main body of a free-standing wall, for example, but not for the top course which will bear the brunt of the weather. Similarly, they should not be used for building earth-retaining walls, or for brickwork underground. For jobs like that you need the extra weather resistance of *special quality* bricks; these are very dense, they do not absorb water, and are thus very durable. Stronger still are *engineering bricks*, which are sometimes used to form a damp-proof course (DPC) in walls; you are unlikely to need these for garden projects.

Unfortunately, the quality is not marked on individual bricks, only on packs or pallets, so when you are ordering small quantities you will have to ensure that you tell your supplier what you will be using them for. If you are using large quantities, look for the following markings on outer wrappings: 0 means zero frost resistance (i.e. internal quality) M and F moderate and good frost resistance.

Type This describes the appearance of the brick. For general work, two types are widely used – *commons* and *facing* bricks. The former are just that; bricks you can use anywhere where appearance does not matter. Facing bricks have a decorative finish on the brick faces that will be exposed to the weather when they are built into a wall, and they come in a huge range of colours and textures. They will be your likeliest choice for most garden projects.

There are two other terms you may come across when shopping for bricks. The first is *stock* brick, which describes the most commonly available brick in a particular district (such as the yellow Kentish brick known as the London stock). The term is

Fig 4 (*below*) The different bricks you are likely to use for general bricklaying projects include faced bricks (*below left*), commons (*centre*) and facing bricks (*right*). Faced bricks have one 'good' face and end which is laid exposed, while facing bricks have good faces and ends all round. Commons are best reserved for out-of-sight brickwork.

Raw Materials

virtually meaningless now that bricks made in one area can be easily transported to any other. The second is *fletton*, which describes clay bricks (commons or facings), made from shale found around Fletton (near Peterborough).

Shape   This takes us into the realms of special bricks – in other words, bricks in shapes other than the standard oblong. Some are designed for use as wall cappings, (the exposed top course of a wall), and come in half-round, bull-nose, saddleback and other shapes. Then there are corner bricks for creating rounded internal or external corners and for finishing off capping runs, rounded end bricks, and so-called plinth bricks which have one top edge cut off at an angle of 45° and which are used to step back the face of a wall neatly where the thickness is reduced above a plinth. All come in a range of colours and finishes, and are either ordinary or special quality (cappings are always special quality). You generally have to order them as few merchants carry large stocks of special bricks.

Size   This is unlikely to bother you much if you are using bricks only for garden projects, but for the record a standard brick measures 215 × 102.5 × 65mm (8½ × 4 × 2⅝in). So allowing for a standard 10mm (⅜in) thick mortar joint between adjacent bricks, the 'work' size is 225mm (8⅞in) long and 75mm (3in) high. Use these figures when you are calculating how many bricks you will need to build a wall of a given height and length (the answer is sixty bricks per square metre for a wall 102.5mm thick, and 120 for one 215mm thick).

Walling Blocks

Man-made walling blocks are available in a range of imitations of natural stone, with the faces either weathered and smooth or textured to resemble natural split stone. They are made from reconstituted stone, (concrete mixed with crushed rock aggregate, often with added pigments to match the natural colour of sandstone or limestone), and are hydraulically pressed to give them strength. Colours range from shades of grey through the buffs to stronger sandstone hues.

Single blocks are widely available in a coordinated range of sizes so they can be laid in imitation of traditional stonework bonding patterns, with a mixture of large and small stones. They have flat tops and bottoms, and so are easy to lay in even, regular courses just like brickwork. Sizes vary from manufacturer to manufacturer, so it is wise to check the precise dimensions once you have found a block you like the look of. A typical standard single-height block measures 230 or 300mm (9 or 12in) long, 100 or 150mm (4 or 6in) wide and 65mm (2⅝in) high.

You can also buy larger blocks which have their outer face moulded to resemble a number of smaller, randomly-shaped blocks, rather like a drystone wall. The joints between the individual 'stones' of these blocks are deeply recessed, so when they are laid the actual joints between the blocks need recessing to match, and should be made with a mortar that closely matches the colour of the block if your little deception is to go unnoticed! Both types come with matching coping stones.

Fig 5   Engineering bricks (*below left*) are used where you need great strength and resistance to water penetration – door steps, for example, or the bottom two courses of a garden wall. Calcium silicate bricks (*centre*) are made from a mixture of sand or flint and lime, and are much more uniform on colour and texture than clay bricks. They come in a wide range of colours. Concrete bricks (*right*) more closely resemble clay bricks, and come in many different colours and textures.

Screen walling blocks These are a variation on the man-made walling theme. They are square, and are pierced with a variety of simple designs, allowing you to build them up into an attractive see-through screen wall. Since they are stack-bonded in columns, they need the support of piers at regular intervals, either of brick or block-work, or built with special pier and corner blocks which have a hollow core, allowing the columns to be reinforced for extra strength – essential for walls built up to more than two blocks in height.

The blocks are generally 300mm (12in) square and 90mm (3½in) thick, with pier blocks 200mm (8in) tall so three pier or corner blocks match the height of two wall blocks. They are generally white or off-white and come with matching coping stones and pier caps.

Fig 6 (*above*) Decorative garden walling blocks come in an immense variety of shapes, sizes, textures and colours, ranging from small brick-size units to large multiple stones. Most are sized in multiples of the standard brick height (65mm) to allow for even coursing, and have matching coping stones and pier caps.

Fig 7 (*left*) Decorative screen walling blocks come in a number of pierced designs, and each range has matching pilaster blocks, coping stones and pier caps.

Paving Slabs

Almost all the paving slabs sold for DIY use are made from reconstituted stone, as used for walling blocks. The cheapest types are simply cast in a mould, and are rather brittle as a result; more expensive slabs are hydraulically pressed for extra strength, and often have a surface texture. This may be random and designed to resemble natural materials such as York stone or slate, or a regular pattern intended to imitate granite setts, herringbone brickwork or interlocking quarry tiles. There is a wide range of colours and surface textures available; squares and rectangles are by far the commonest shapes, but you can also buy hexagons, circular stepping-stones and slabs with quadrant cut-outs for paving round trees and other obstacles.

Sizes for regular slabs range from 225 or 300mm (9 or 12in) square up to 600mm (24in) square, with rectangles from 450 × 225mm (18 × 9in) up to 675 × 450mm (27 × 18in). Hexagons and circles are usually about 400mm (16in) across.

Paving Blocks

These are small concrete blocks resembling bricks, designed to be laid in interlocking patterns on a sand bed between fixed edge restraints. They come in a wide range of colours and textures and are mostly rectangular, measuring 200 × 100mm (8 × 4in) and 65mm (2½in) thick. Interlocking shapes are available too.

Fig 8 (*above*) Most of the paving slabs available are made from reconstituted stone, and basic squares and rectangles come in dozens of colours and a range of different surface textures.

Fig 9 (*left*) You can also buy paving slabs with pressed finishes that imitate materials such as granite setts, herringbone brick work or mosaic tiles. Round stepping stones let you lay a path across the lawn, while interlocking hexagons provide an alternative to a square grid layout.

Cement

Cement is the basic ingredient of both mortar (for bricklaying) and concrete; it is actually the adhesive that binds them both together. It is made from a mixture of lime and clay, ground to a powder, and sets when mixed with water and so needs to be kept perfectly dry if it is stored for any length of time (in fact, most cement now has a sell-by date, and supplies more than a few weeks old should be thrown away). Any cement that has become damp and has formed lumps should also be disposed of since it will not set properly even if the lumps are broken up.

Ordinary *Portland cement* (commonly referred to as OPC) is the familiar grey powder which is the least expensive and most widely used type. White Portland cement is identical except for its colour, and is used for jobs demanding a white mortar or concrete finish – it is roughly twice the price of OPC. If you live in an area with clay soil, it may be worth asking your local authority's Building Control Officer if you should use sulphate-resisting cement for constructional projects in contact with the soil.

Masonry cement is Portland cement with special additives (*see opposite*) that increase its plasticity, and it is used for making bricklaying and rendering mortars – it needs only the addition of sand to make a good mortar. It should not be used for making concrete.

Aggregates

The other major ingredient of mortar and concrete mixes are the aggregates which the cement binds together.

Sand is used to make both mortar and concrete, and contains particles no bigger than will pass through a 5mm sieve. For bricklaying and rendering, you need so-called soft sand (also known as builders' or bricklayers' sand). This contains a high proportion of fine particles, which helps to create a good, workable mortar. For concreting, you use sharp concreting sand, which contains coarser particles. The colour of sand varies widely around the country, and is an important factor which affects the look of the mortar or concrete into which it is mixed.

Coarse aggregates are sieved stones (not chippings) sold in various size ranges

(5 to 10mm, 10 to 20mm, 20 to 40mm, for example) for use along with sharp sand in making concrete; the larger the stones, the coarser the mix. You can also buy combined sand and coarse aggregate, known as *all-in aggregates* or ballast; these are ideal for general concreting work, but as the proportions of sand to stones in naturally-dug ballast can vary it is better to mix separate ingredients for jobs where the strength of the concrete is important – in particular for foundations.

Fig 10 (*above*) The basic ingredients for mortar are Portland cement (*foreground*) and soft bricklaying sand. For concrete, sharp sand is used instead, with the addition of small graded stones called aggregate.

Raw Materials

Additives

The main additive used in mortar and concrete is a *plasticiser*, which increases the workability of the mix and assists in retaining water. Lime is the traditional plasticiser, and helps to reduce the strength of pure cement/sand mixes which would otherwise be too strong for general use. You can also buy liquid plasticisers which are added to the mix water.

Other additives which you may need to use for particular jobs are *pigments* for producing coloured mortar or concrete, *waterproofing agents* to improve their weathering properties, and *frost-proofers* which are used when working in cold weather. See page 90 for more details.

Plaster

Plaster for indoor use is based on gypsum, with other minerals sometimes added to make the plaster lighter in weight. It is usually applied as a two-coat system, with a relatively thick base coat or undercoat followed by a thinner finish coat.

Carlite plasters are the most widely available lightweight brand in the UK. The type to use on normal backgrounds such as bricks and lightweight building blocks is called Browning plaster; there is a special version called Browning HSB for use on very porous walls, where ordinary Browning plaster would crack. For plastering on dense non-absorbent surfaces such as concrete, you should use Carlite Bonding plaster instead. Carlite Finish plaster provides the top coat. For plastering over expanded metal mesh (when using arch formers, for example), you need a special grade called Metal Lathing plaster.

Thistle plasters are non-lightweight gypsum plasters. Thistle Browning is used, mixed with sharp sand, as an undercoat on normal backgrounds, with Thistle finish providing the finish coat. Thistle Board Finish is used on plasterboard.

Sirapite is another non-lightweight plaster, used mainly as a finish coat over cement/sand undercoats, and for repairs.

To cut down on the time needed to apply two-coat plaster systems, especially on very uneven backgrounds, you can buy so-called one-coat plasters. These can be applied in thicknesses of up to 50mm (2in) – four times the thickness of a normal two-coat system.

Lastly, if you are afraid of using traditional plasters you can buy small tubs of ready-mixed undercoat and finishing plaster which are ideal for patching work, but expensive for large areas.

Fig 11 (*above*) Additives for mortar and concrete mixes include liquid plasticiser, waterproofer and frostproofer, while powder pigments allow you to create coloured mixes.

Fig 12 (*far left*) Gypsum plasters of all types are sold in sacks ready for mixing. Sirapite is used only as a finish plaster, and is popular with amateur plasterers because of its relatively slow setting time.

Fig 13 (*left*) Ready-mixed undercoat and finish plasters are ideal for beginners but expensive for large areas.

Tools and Equipment

You will need quite a range of tools to carry out the sort of constructional and repair jobs dealt with in this book. Some of the more general-purpose tools such as a club hammer, a spirit level or a garden spade you will probably have already; other more specialist items will have to be bought as and when you need them. Some, expensive and seldom used, are better hired than bought, see page 19.

Mixing Tools

The first step for many building projects is mixing up raw materials for mortar, concrete or plaster. You can mix plaster and small quantities of mortar and con-crete using nothing more sophisticated than a plastic bucket and a piece of wood, but for larger mixes you need a hard flat surface to work on, a common-or-garden spade to do the mixing and a bucket or watering-can for the mixing water. You can use your drive, path or patio for mixing mortar and concrete so long as you hose away the remains as soon as you have mixed and moved the pile to avoid staining the surface. If this is a problem, or you have not got a suitable area, you can buy strong plastic mixing trays about 1m (3ft) square from DIY superstores. They have a raised lip and can hold one 40kg bag of dry ready-mixed mortar, for example. Otherwise, use an old sheet of plywood.

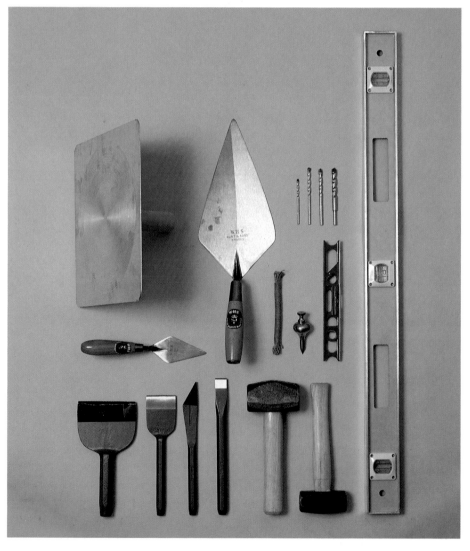

Fig 14 (*left*)   The tools you will need for bricklaying include: hawk, pointing trowel, bricklaying trowel, stringline and pins, spirit levels (*top row*); brick bolster, cold chisels and club hammer (*bottom row*).

Tools and Equipment

Bricklaying Tools

The key tool for laying bricks and walling blocks, and any other job involving using mortar, is the *bricklayer's trowel*. This has a blade between 200 and 300mm (8 to 12in) long – large enough to deliver a reasonable amount of mortar in one go, yet not too heavy to handle. Its wooden or plastic handle is extremely useful for tamping bricks or blocks down into position once they have been laid. There are several designs, but the London pattern is the commonest. This has a cranked handle and a kite-shaped blade with a slightly rounded tip.

The *pointing trowel* is its smaller relation, and is used as its name implies for pointing or finishing the surface of the mortar layers (called courses) between the bricks. The trowel blade is anything from 75 to 200mm (3 to 8in) long, and has a sharper point than the bricklaying trowel so it can reach into tight corners. It is a useful tool to have in your box for all sorts of small-scale repair jobs, as well as for pointing.

Also worth investing in, especially for repair work, is a *hawk* – a small square metal or plastic board with a handle attached to one face, used for carrying small quantities of mortar or plaster to wherever you are working.

There are two other tools you will need for laying bricks – one useful if you want to do things properly, and one essential. The first is a pair of *bricklayer's pins* and a *stringline*, used as a guide to ensure that the bricks in each course are laid level and in line with each other. Once the ends of the wall have been started, the pins are inserted into the mortar joints at each end of the wall and the stringline is stretched between them, see page 23. The second is a *spirit level*, which you need to check that your work is built to a true vertical and horizontal. On small scale jobs it can take the place of pins and stringline. You can get by with a small spirit level used in conjunction with a long timber straight-edge, but if you intend to do more than the odd repair job it is well worth investing in a long lightweight aluminium builder's level with vials to indicate true verticals as well as true horizontals.

You are bound to have to cut bricks, blocks and paving slabs at some stage during your bricklaying career, and the traditional tools for doing this are the *brick bolster* and *club hammer*. The bolster is a type of cold chisel, and has a hexagonal shaft squared off at one end to form the hammer anvil (the end you hit), and flattened out at the other into a spade-shaped blade between 75 and 100mm (3 to 4in) wide. The tool is about 185mm (7½in) long overall, and is often sold with a plastic guard round the shaft to protect the hands from misguided hammer blows. The club hammer used to drive the chisel has a squared-off head weighing up to 1.8kg (4lb), attached to a short wooden shaft about 300mm (12in) long.

You will need some smaller *cold chisels* too, for cutting holes in walls, chiselling out plaster and repairing old pointing. They come in a wide range of sizes; most are square-ended like the brick bolster, but one, the plugging chisel, has an angled blade and a sharp point. It was designed for chopping holes in masonry for wooden plugs that provided fixing points for things like window frames (a job now done by wallplugs of various types) but is useful in hacking out old pointing.

Fig 15 (*far left*) A builder's square is an essential aid for setting out foundations and for ensuring that brickwork is built up square. Make one from scrap timber with sides in the ratio of 3:4:5 to form a perfect right-angled triangle. The optimum size is 450:600: 750mm. Join the two shorter lengths with a halving joint, then add the longest piece to form a rigid structure.

Fig 16 (*left*) A gauge rod is used to check that your brick courses are evenly spaced. It is a length of straight timber marked off along its length to represent 65mm high bricks separated by 10mm thick mortar courses.

Tools and Equipment

For making fixings into masonry, you will need your power drill plus a range of *masonry drill bits*. These have specially hardened tips, and come in a wide range of diameters and lengths.

That just about completes your tool requirements, at least for everyday building projects. You may need sundry things like a wheelbarrow for moving materials around, a stout timber plank for tamping concrete down and so on, and there are a couple of work aids you can make yourself. The first is a large *builder's square*, made by fixing together three lengths of wood into a right-angled triangle. The second is a *gauge rod* – a length of wood marked off along its length to represent 65mm (2⅝in) wide bricks separated by 10mm (⅜in) wide mortar joints, and used to check that your walls are building up evenly.

Plastering Tools

If you plan to tackle plastering – or at any rate, plaster repairs – you will need a tool called a *float* or plasterer's trowel. This has a rectangular steel blade measuring about 250 × 115mm (10 × 4½in) and a shaped wood or plastic handle, and is used for applying plaster to the wall and finishing it off smoothly. You will also need a hawk for holding small quantities of plaster close to the wall as you work, a flat piece of board (known as a spot board) for holding the mixed plaster, and lengths of scrap wood for use as plastering rules and guides.

Optional extras for plastering work include special *corner trowels* with folded blades for finishing internal and external corners, and small pointing and mastic trowels for coaxing plaster into awkward or confined spaces.

Access Equipment

Last but not least, you are sure to need steps, a ladder, trestles, perhaps even a hired platform tower to enable you to work comfortably or reach inaccessible surfaces in and around the house. Make sure such equipment is in good order, properly erected and used safely, *see* page 19.

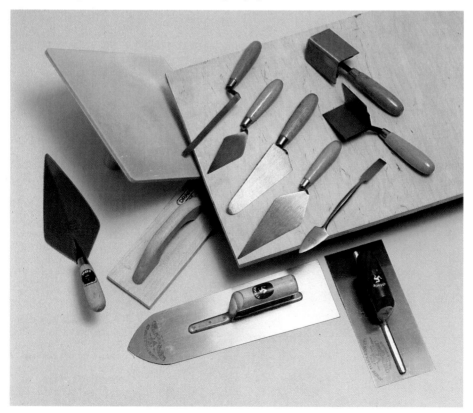

Fig 17 (*left*) The basic tools for plastering are the rectangular steel float (*bottom right*) and the hawk (*top left*). The wooden float is mainly used as a devilling float for keying undercoats, while the boat-shaped float is used for spreading thin screeds on floors. The tools on the spot board include internal and external corner trowels and a range of pointing and mastic trowels used for getting plaster into awkward corners where the steel float cannot reach.

Shopping and Safety

You will need quite a range of tools and materials for the various building and maintenance projects featured in this book. Where you shop for them depends on the scale and nature of the job; for a simple repair you may need to look no further than your local hardware store, but for more extensive projects you will make big savings by shopping around for materials. Here are the places to try.

Local DIY Shops

The typical independent high street DIY shop usually stocks things like fillers, small bags of dry ready-mixed mortar or patching plaster, flashing and other repair tapes, but is highly unlikely to stock any other building materials. You will probably be offered a choice of just one or two brands. You should also be able to buy any general-purpose tools you need but again with little brand choice.

Verdict Fine for small jobs if it has what you need and the convenience outweighs the disadvantage of highish prices. Generally good at offering helpful advice, and a useful source of contact for local contractors for large-scale jobs.

DIY Superstores

The major national chains all offer a good range of materials for building projects (bricks in one or two varieties, garden walling blocks and paving slabs, for example) and some will arrange local delivery so you won't have to wreck you car's shock absorbers getting a load home, see Tip. They also stock dry ready-mixed mortar and concrete, and some sell bagged cement, sand and aggregates separately. You will also find plaster, textured finishes and the like. They should stock all the tools you will need too. All sell ladders and steps, and some operate hire sections which can be useful for access equipment and specialist tools, see Hiring Things.

Verdict Good range of relevant products (with the one exception of pure building supplies) usually at reasonable prices, and convenient opening hours.

Builders' Merchants

A good source of supply for 'heavy' building materials like bricks, garden walling blocks, paving blocks and slabs. Many stock a wide range of different bricks, or can obtain them for you if your order is large enough. They will also have the biggest choice of other building materials and will deliver large quantities of things like cement, sand and aggregate too. Some have 'retail' counters designed for the non-trade customer.

Verdict Wide selection of goods at reasonable prices. Good for large-scale projects and will always deliver, but usually closed at weekends.

Garden Centres

Many garden centres are now stocking a good range of garden walling and paving materials, plus associated products such as pond liners, bird baths and even garden statues! Some will also arrange local delivery. However, do not expect to find raw materials like bricks, cement and sand – you will have to shop elsewhere for these.

Verdict Good selection of purely garden building products, but likely to be more expensive than 'trade' outlets. The one big advantage many offer is Sunday opening.

Hiring Things

There is a hire shop in virtually every high street these days – either a branch of one of the large national chains such as Hire Service Shops, or a local firm. Most carry a range of equipment that can be extremely useful for the amateur builder who may need a specialist (and expensive) piece of equipment for just one project. The range stocked varies, but will probably include access equipment of all types, cement mixers, plate vibrators and rollers, block splitters, high-pressure washers, concrete breakers (vital for digging up old drives, paths and patios), even damp-course injection machines. You can also hire power tools such as angle grinders. Go for firms with Hire Association Europe membership, and make sure you are shown how to operate anything you are unfamiliar with.

One specialist outlet you might want to visit, for inspiration if not for the end product, is the fireplace shop. These outlets usually operate on a supply-and-fix basis – but some can be persuaded to supply surround kits for you to install.

SAFETY
There are several dangers involved in building work. The first is the risk of back injury as a result of trying to lift heavy objects like paving slabs, and hand injury from handling coarse or corrosive materials.

Always lift heavy objects with a straight back, and don't think it is sissy to use a barrow or trolley to move heavy things around the site. Wear stout gloves when handling coarse materials, and take care to avoid cement burns (especially if raw cement falls inside your wellies). Wear stout boots to guard your feet against crushing injuries.

Another danger is the risk of falling from access equipment, indoors as well as outside; you can have quite a nasty fall stepping back off a low trestle, for example. Make sure that steps, ladders, platform towers and trestle supports are set up square and level on solid ground, and tie ladders and towers to the building with rope for extra safety.

TIP
Do not try to load more than a few bricks or a bag of cement in your car boot; a heavy load will affect the way the car handles, and could seriously damage the suspension. Either buy from firms offering a delivery service, or 'hire' a driver with a van or lorry to collect the goods for you.

Mixing Mortar and Concrete

Mortar is a mixture of cement, soft sand and either lime or a chemical plasticiser to which you add water to make a self-hardening paste. Concrete is similar, but consists of cement, sharp sand and aggregate (lime and plasticisers are not used in concrete mixes). In both cases, the secret of success lies firstly in measuring out the quantities you need accurately – this is just as important as it is in cooking if you want good results. The second essential is thorough mixing of the ingredients to produce a uniform mix of even moistness and consistency. For small jobs you can do this quite adequately by hand; for large-scale concreting projects it may be worth considering hiring a concrete mixer to take the hard work out of the job, or even buying in ready-mixed concrete from a local supplier.

Cement, lime, sand and aggregate are sold in bags of various size up to 50kg (about 1cwt); sand and aggregate are also sold loose in bulk and are delivered by lorry. For small jobs you can buy bags of dry ready-mixed mortar and concrete.

What to do

Start by measuring out the dry ingredients by volume, using a stout plastic bucket or similar container. Choose the mix formula you need for the job in hand – see Check – and pour the ingredients out on a smooth, hard surface. Use a chipboard or plywood offcut or a plastic mixing tray if you have no suitable surface. Then turn the mix over with your shovel to mix the ingredients dry, before starting to add the water. Stir any pigments or other additives you are using into the mixing water first, then form a crater in the middle of the heap and pour in half a bucket of water to start with. Mix in the dry ingredients from the edges of the heap, turning over the whole mix so the ingredients blend thoroughly. Add more water gradually until the mix is stiff enough to retain ridges formed with the shovel. If it is too sloppy the mix will be weakened; add more ingredients, correctly proportioned as before, to stiffen it up again. Do not mix more mortar or concrete than you can use within an hour or so.

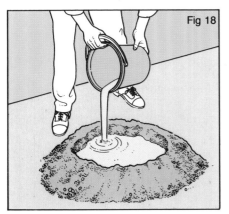

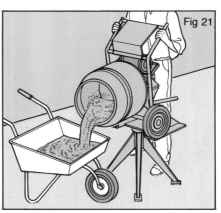

Fig 18   Dry-mix all the ingredients thoroughly, then form a crater in the middle and add half a bucket of water.

Fig 19   Turn the whole pile over until it has a uniform consistency that will retain ridges.

Fig 20   Mix very small quantities in a bucket.

Fig 21   Use a concrete mixer for large jobs. Add half the aggregate first, then half the water, then the cement and the rest of the aggregate, and add more water as necessary. The mix should fall cleanly off the mixer blades.

Setting Out for Bricklaying

Any bricklaying project, large or small, needs some planning if it is to be a success. In other words, you need to set out some form of guide-lines to show you where the bricks will actually be placed as you lay them. This also allows you to check that corners will be square, and that the finished project will be precisely the size and shape you intended it to be, by allowing you the opportunity to set out the first couple of courses as a dry run.

On a simple project you can set out these guide-lines using string and bricks, or do the job the way the professionals do using what are called profile boards – two pegs with a horizontal crosspiece which can mark the size of the foundations if you are laying them as well as indicating the position of the brickwork. The advantage of using profile boards is that they can stay in place with the stringlines temporarily removed while work progresses; all you have to do to check that things are proceeding accurately is to wind the strings back round the nails in the top board. Always double-check angles and dimensions carefully at this stage.

What to do

Let us assume at this stage that you are setting out your first project on an existing hard surface – a concrete patio, for example, on which you want to build a small brick planter or a barbecue. This will serve as an adequate foundation here; see pages 26–8 for how to lay foundation strips and slabs for larger projects.

On a solid base you will have to use bricks and stringlines. Tie the string round one brick and set it on the slab just beyond the edge of the project work area. Then secure the other end round a second brick and pull the string taut. Use more bricks and string to set out the positions of any corners, and check that these are at right angles using your builder's square. Then dry-lay your first two courses of bricks to check the dimensions and if possible to minimize unnecessary cutting.

With profile boards, used in conjunction with freshly-laid foundations, set out the pegs beyond the foundations and then attach the stringlines to nails driven into the crosspieces.

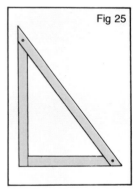

Fig 25

Fig 25   Builder's square.

Fig 22

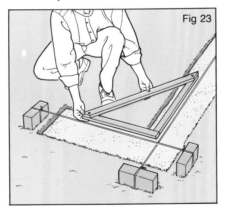

Fig 23

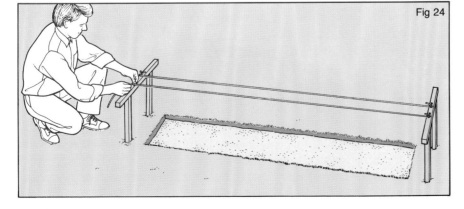

Fig 24

Fig 22   Tie your stringlines round bricks and pull them taut to mark the outline of your projected structure.

Fig 23   Use your builder's square to check that all corners are true right angles. Adjust the positions of the stringlines if necessary.

Fig 24   Use profile boards with newly-laid foundations, driving the pegs in outside the work area and running strings between nails driven into the crosspieces.

Laying the First Course

Once you have set out your stringlines and carried out a dry run to check the brick layout, you can mix up your first batch of mortar (see page 20) and prepare to lay your first bricks.

Your first projects are likely to involve building structures with walls one brick thick, and to give extra strength the individual courses are laid with each brick overlapping two bricks in the course below it. This simple overlapping pattern is known as *stretcher bond* because it exposes only the side faces of the bricks (the stretchers) when the wall is complete. More complex bonds (see page 25) are used for thicker (and taller) walls.

Assuming that your first course consists entirely of whole bricks, you will have to use some half-bricks in the second course to maintain the bonding pattern. On a straight run of wall, for example, the second course (and all subsequent even-numbered ones) will start and finish with a half-brick, the rest of the course being laid with whole bricks as before so each is offset by half its length. The third course is the same as the first.

What to do

The technique involves scooping up a slice of mortar with your bricklaying trowel, depositing it on the base or foundation beneath the stringlines and spreading it out roughly with the end of the trowel so it is slightly wider than the bricks themselves. You then place the first brick in position and tamp it down into the mortar bed so it looks level (you can check and adjust the level more precisely when you have laid several bricks).

Next, use your trowel to butter some mortar onto the end of the next brick and set it in place, butting it up against its neighbour so the mortar between them is compressed to a thickness of about 10mm (⅜in). Again tamp it down so it looks level and in line. Repeat this procedure until you have laid about five bricks, then use your spirit level to check that all are sitting level and in line with each other.

Carry on laying bricks in this way along the line of the course, checking line and level every so often, until you reach the end of the first course.

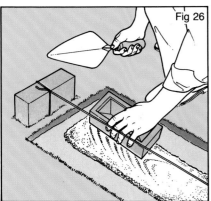

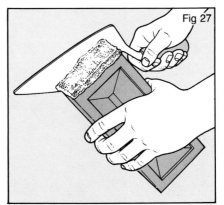

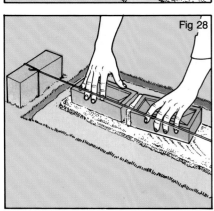

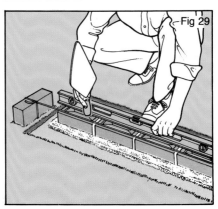

Fig 26 Spread mortar in a line on your base or foundation strip, and position the first brick. If you are working with profile boards, use a spirit level held vertically to mark the position of the first brick on the mortar directly below the stringlines.

Fig 27 Butter some mortar onto the end of the next brick.

Fig 28 Lay the second brick in line with the first, butting it up so the mortar between them is compressed to about 10mm (⅜in). Tamp it down into the mortar bed as before.

Fig 29 Lay four or five bricks, then use your spirit level to ensure that all are level and in line. Tamp bricks down as necessary with the trowel handle, and raise low ones by adding more mortar.

Building Up the Wall

Once the first course is laid level and true, you can start building the wall up course by course to its designed height – around seven or eight courses is the maximum recommended height for a single-brick wall with no additional reinforcement in the form of piers (see page 24) and is also a suitable height for structures such as low-level planters.

To ensure that cumulative errors such as over-thick mortar joints do not occur, you build up your wall by working from the ends towards the middle. If you simply started at one end, you might reach the far end of the second course and find that there was not enough room for the last brick to be positioned.

You will find that as you become more adept at bricklaying, you will be able to gauge the thickness of the horizontal joints by eye, but until then it helps to use what is known as a gauge rod – a length of wood marked off with 65mm (2½in) brick thicknesses and 10mm (½in) joint thicknesses, and used to check that the courses are rising evenly as you build them up to the finished height.

What to do

Start at one end of the wall by spreading some mortar on top of the first course, and then bed the first brick in place (use a half-brick if you started with a whole one in the first course). Then lay as many whole bricks along the wall as there will be courses in the wall. Do the same at the other end of the wall.

Next, add bricks to start the third, fourth and subsequent courses, until you have reached the final wall height at the very end of the wall. The bricks form even steps half a brick long running down to the level of the first course. Repeat the process at the other end, using your gauge rod to check the joint thicknesses.

Now you can use your bricklayer's pins and line to run a stringline along the wall. Push the pins into the mortar joint between the second and third courses and stretch the line taut. Then you can fill in the missing bricks in the second course. Move the line up one course at a time and repeat the process until you reach the top course, which you can lay without a line.

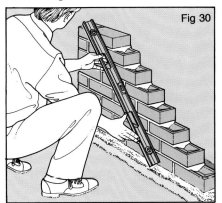

Fig 30 Build up stepped courses at each end of the wall until you reach the topmost course, and check that there are no bulges or hollows by holding a straight-edge or spirit level against the face of the wall.

Fig 31 Use your gauge rod at each end of the wall to check that your coursing is even.

Fig 32 Set your bricklayer's pins and line in place and fill in each course in turn with whole bricks. Move the pins up as you proceed. Lay the top course of bricks with the hollows facing downwards.

Forming Corners and Piers

Unless you intend to spend your bricklaying career building short straight lengths of wall, you will next need to master two new techniques: how to build corners, and how to reinforce your walls by adding supporting piers at intervals.

So long as you are working in stretcher bond, turning corners could not be simpler. All you do is place each corner brick at right angles to the one beneath it; this automatically maintains the bond as the wall turns the corner, and also effectively ties the two adjacent sections of wall strongly together.

As for piers, these too are bonded into the wall structure for maximum strength; it is not enough simply to stack up a pile of bricks on top of each other against the face of the wall. For free-standing walls laid in stretcher bond, you need piers at each end of the wall and every 3m (10ft) along it if the wall height exceeds 450mm (18in). Even with piers, such a wall should not exceed about 675mm (27in) in height; anything taller should be built in brickwork 215mm (8½in or one brick length) thick.

What to do

To turn a corner in stretcher bond brickwork, lay the entire first course to start with, turning the corner by laying a whole brick at right angles to the first section laid. Then form the corner of the next course with two bricks laid in the opposite direction to those in the first course and build the corner up like the wall ends (see page 23) before filling in the rest of each course as before.

To build an end pier, lay one brick at right angles to the last wall brick in the first course, and fill in the angle between them with a half-brick. Start the next course of the pier with two bricks laid at right angles to the one in the first course. Subsequent courses simply repeat the layouts of courses one and two.

For an intermediate pier, lay two bricks side by side at right angles to the wall face in the first course. In the next course, to avoid vertical joints aligning in adjacent courses, you lay two threequarter-bricks and a half-brick on the wall face of the pier and a whole brick at the back.

What you need:
- bricks
- mortar
- bricklaying trowel
- spirit level
- gauge rod
- bricklayer's pins and stringline
- builder's square
- club hammer and brick bolster

Fig 37   Use a brick bolster and club hammer to cut ½ or ¾ bricks.

Fig 33   How the bricks are laid in alternate courses to form an intermediate pier. Note the ½-brick and two ¾-bricks in one of the two repeating layouts.

Fig 34   How the bricks are laid to form an end pier. There is a ½-brick in every other course.

Fig 35   How the bricks are laid to turn a corner in stretcher bond.

Fig 36   For a neat finish to the mortar, draw the edge of your trowel along each joint with its underside resting on the top edge of the brick below.

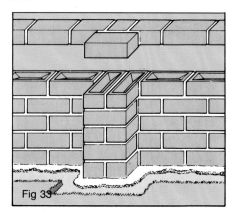

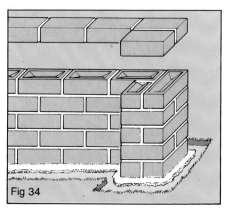

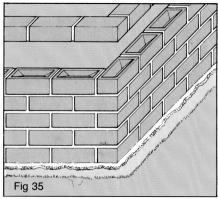

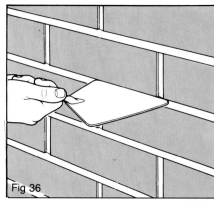

Pointing and Other Bonds

The final stage of building a brick wall or similar structure is the pointing, giving the exposed edges of the mortar joints a neat, weatherproof finish so that rainwater cannot penetrate behind the face of the brick and freeze, causing the characteristic splitting of the brick face known as spalling. You can do this in several ways, and it is easiest for the amateur to tackle the job after completing each area of wall.

What to do

The simplest joint to form is called a weathered joint, illustrated on page 24. Alternatives include the profiled joint, made by drawing the tip of the trowel along each joint to leave a V-shaped recess, and the concave joint formed by drawing an object such as a piece of garden hose along the joint. Better in exposed areas is a variation on the weathered joint that leaves a slight overhang along the bottom edge of the joint. Form it as for a weather-struck joint, leaving a slight mortar overhang, which should be trimmed off neatly with a trowel and straight-edge.

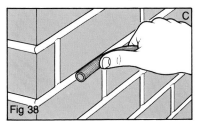

Fig 38

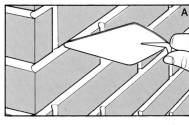

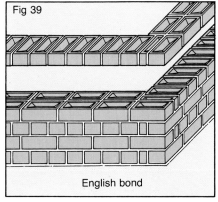

English bond

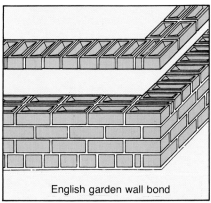

English garden wall bond

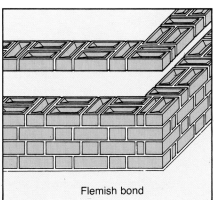

Flemish bond

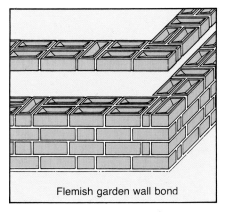

Flemish garden wall bond

Fig 39

What you need:
- pointing trowel
- timber straight-edge
- piece of hose or similar rounded tool

Fig 38   Finish the mortar joints with a weathered joint (*page 24*), or use a V-joint (*A*), an overhanging weathered joint (*B*) or a concave joint (*C*).

Fig 39   For walls over about 675mm (27in) high you need the extra strength of brickwork 215mm (one brick length) thick. This allows you to use one of several different bonding patterns as you build up the courses.

English bond is laid with the first course containing two rows of parallel stretchers, followed by a course of headers (bricks laid end-on to the face of the wall). At corners a brick split lengthways (a queen closer) is used to maintain the bond.

English garden wall bond has from three to five courses laid as stretchers, followed by a course of headers as in English bond.

Flemish bond has each course formed by laying a pair of stretchers followed by a header, again with queen closers at corners and ends to maintain the bond.

Flemish garden wall bond is a variation using one header after about three pairs of stretchers in each course.

Setting Out for Concreting

Concrete is one of the most versatile, and economical, building materials for a wide range of jobs, from laying paths, drives and patios to forming foundations for building projects and sturdy bases for outbuildings. Where you plan to lay a large expanse of concrete, the secret of success lies in accurate planning and careful setting-out of the site.

Start by working out exactly where the slab is to go, and what its size and shape will be. For a simple square or rectangle, you can simply drive timber pegs into the ground to mark the corners, but for more elaborate shapes it is best to do a scale drawing on squared paper first to make it easier to calculate the surface area and so to estimate quantities accurately.

What to do

For a square or rectangle, start by driving in pegs at the four corners to indicate the size of the slab. Then drive in pairs of pegs outside each corner so you can stretch stringlines along each side of the slab. Secure them to nails tapped into the top of each peg, then use your builder's square to ensure that the whole site has right-angled corners. Another way of checking that corners are square is to measure the diagonals; if these are the same you know you have right-angled corners.

The purpose of the pegs outside the work area is to allow you to remove the four original corner pegs and the stringlines so they are out of the way while you excavate the site, while leaving the other pegs in position so you can quickly re-instate the stringlines and check the accuracy of your digging.

To transfer plans for an irregular area to the site itself, use stringlines to divide the site up into squares matching those on your sketch plan, and then drive pegs in or use lengths of garden hose-pipe or thick rope laid out across the site to mark out the perimeter of the area to be concreted. Work from your plan one square at a time, and then view the planned-out area from an upstairs window so you get a better view of its overall shape and can make any adjustments that may be needed to the layout.

What you need:
- 25mm (1in) sq pegs
- mallet
- string
- nails
- builder's square
- steel measuring tape

CHECK
- that corners are square by measuring the site diagonals. They should be equal.
- that pegs are long enough to be driven well into the ground – a length of 300mm (12in) is ideal. Roughly sharpen the points to make them easier to drive in straight.

TIP
Work out the area of squares and rectangles by multiplying length and width. For irregular shapes, draw a scale plan on squared paper so you can estimate how many whole and part squares the design will cover.

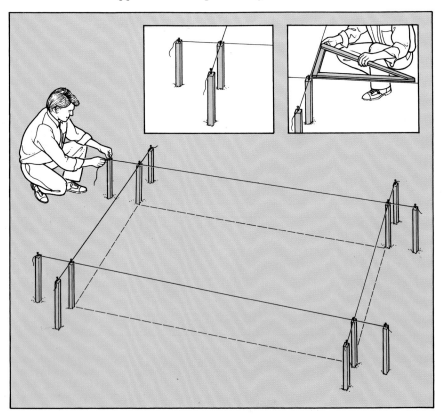

Fig 40 Mark out the site for a square or rectangle of concrete by driving pegs in to mark the positions of the corners. Check the dimensions, then drive in two more pegs at each corner, outside the site area, so you can attach stringlines to run over the tops of the corner posts. Check that all corners are square, then you can remove the inner corner pegs and the stringlines and excavate the site without tripping over pegs or string. Simply replace the stringlines as necessary to check your progress.

Using Formwork

If you are laying a strip foundation for a garden wall, you can normally rely on the sides of the trench to contain the concrete as you place it. After all, the strip will be out of sight once the wall is finished, so all that matters is that the surface of the strip is level, see page 28. However, slabs cast as areas of hard standing or as the base for an outbuilding look and perform better if they have neat vertical edges all round. The way to provide these is to lay your concrete within what is known as formwork or shuttering – lengths of timber that form a mould for the concrete. They are removed once the concrete has set.

Formwork does not just contain the concrete, however. If it is accurately set out, it also makes it easier to smooth and level the surface of the slab, since the edges of the mould also act as a levelling guide. You can use any timber that is wide enough to form the mould; old floorboards are ideal, but if you have to buy wood it is cheaper to use strips of shuttering-quality plywood rather than softwood. For large-scale projects, it is cheaper to hire the formwork.

What to do

Once you have excavated the site for the slab to the correct depth (see Check), drive 50mm (2in) square pegs into the ground all round the perimeter to support the formwork boards. They should be about 1m (3ft) apart, and must be set to the correct level. Then you can nail the formwork in place to the pegs, butting the end of one board against the face of the next one at corners. Again check that the board edges are level. If you have to butt-joint boards along the edges of the formwork, add extra pegs to support the joint. If the slab needs a slight fall across it for drainage, adjust the peg and board positions accordingly to give a fall of about 1 in 40 from one side of the slab to the other. Once all the formwork is in position, check that corners are square and that the overall dimensions are correct.

You can also use formwork to lay slabs with curved edges. Use plywood strips that are thin enough to be bent to the desired radius, and hold them in place with more closely-spaced pegs.

What you need:
- 50mm (2in) sq pegs
- mallet
- timber or plywood for formwork
- nails
- hammer
- spirit level
- builder's square
- steel tape-measure

CHECK
- that your excavation is deep enough to accommodate both the concrete thickness and any hardcore needed to stabilize poorly-compacted subsoil. As a general guide, lay slabs 100mm (4in) thick. You can use a 75mm (3in) thick slab for areas up to 2m (6ft 6in) square and subject to light use. For driveways lay a minimum thickness of 150mm (6in).

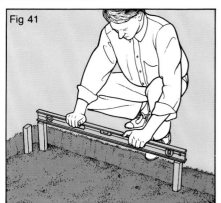

Fig 41

Fig 42

Fig 43

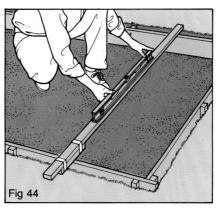

Fig 44

Fig 41 Drive 300mm (12in) long pegs in round the perimeter of the site at 1m (3ft) intervals, and check that their tops are level.

Fig 42 Nail boards to the pegs, bracing them with your other hand as you work so you do not disturb the pegs.

Fig 43 Check that corners are square and that slab dimensions are correct.

Fig 44 If you need a fall across the slab, adjust the peg depths and board positions accordingly. Check the fall by fixing an offcut (25mm thick for each 1m of slab width) to a long straight-edge to give a fall of 1 in 40.

Laying Concrete

The secret of successful concreting lies in using the right mix for the job, mixing it carefully and placing it properly. Assuming that you have got the first two of these steps right (see pages 91 and 20 respectively for more details) you are ready to proceed to stage three.

Before you start, double-check that you have ordered sufficient materials for the project. This is especially important for slabs such as patios, where using sand or aggregate of a different colour or size for part of the job could result in a visibly patchy finish. If you are using the standard general-purpose 1:2:3 mix, the yield per 50kg bag of cement is 0.15cu m – enough concrete to lay 1.5sq m of slab 100mm thick. For strip foundations using the weaker 1:2½:3½ formula, the yield per bag of cement is 0.18cu m – enough for 6m (18ft) of foundation 200mm (8in) wide and 150mm (6in) thick.

If you are using ready-mixed concrete, let the supplier estimate quantities for you based on your plans or specification. Remember that few firms will deliver less than about 3cu m (105cu ft).

What to do

Complete the site preparation by checking that formwork is correctly placed and hardcore (if used) is well compacted. For strip foundations, make sure that datum (level guide) pegs are driven into the foot of the trench at 1m (3ft) intervals with their tops level, as a guide to correct placement of the concrete.

Then mix and load up your first batch of concrete, barrowing it to where it is needed along runs of scaffold boards if you have to cross soft lawn areas. With formwork, tip the concrete in and rake it out, working it well into the corners. Add more loads as necessary until the mould is slightly overfull. Then use a heavy timber tamping beam to compact the concrete with a chopping motion until it is level with the top edges of the formwork.

With strip foundations, either tip the concrete directly into the trench or shovel it into place round the datum pegs. Compact it with a punner (a block of wood nailed to the end of a fencepost) and check that the surface is level with the pegs.

What you need:
- hardcore
- concrete
- wheelbarrow
- scaffold boards
- garden rake
- tamping beam or punner

CHECK
- that concrete is well compacted along edges and into corners when placed within formwork.
- that the formwork is overfilled by about 10mm for every 100mm of slab thickness to allow for thorough compaction.

TIP
If you are laying strip foundations, avoid the risk of the trench edges crumbling by placing a thick plywood sheet across the trench so you can tip up the barrow parallel to the direction of the trench.

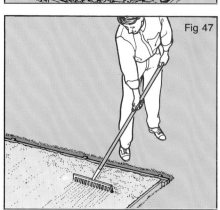

Fig 45

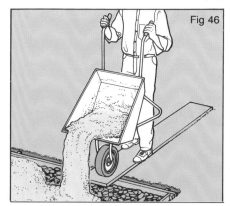

Fig 46

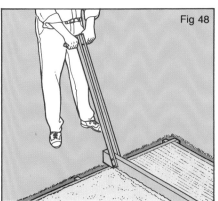

Fig 47

Fig 48

Fig 45 Before starting to place concrete in formwork or in a trench, make sure that hardcore is well compacted.

Fig 46 Wheel barrows to the site on scaffold boards so you can discharge the concrete into the formwork or trench without disturbing formwork or collapsing trench edges.

Fig 47 Rake concrete out evenly, working it up to the formwork along edges and in corners.

Fig 48 Compact the concrete thoroughly using a heavy timber tamping beam with a chopping action until it is level with the top edges of the formwork.

Making Fixings into Masonry

There are plenty of situations where you will need to make fixings into masonry, either into the new brickwork or concrete you have just completed, or into some existing structure. For light-duty fixings you can of course use masonry nails or screws and nylon wallplugs. Think twice before using masonry nails; their ease of fixing is outweighed by the fact that they are almost impossible to remove without damaging the masonry.

For heavier-duty applications – fixing a timber wallplate or an awning to a wall, for example, or securing a timber framework to a concrete base – you need something that will provide a stronger and more secure fixing. The solution for most situations is to use a device called a metal expansion anchor.

This consists of a split metal sleeve containing a threaded cone-shaped insert and an integral bolt. The bolt may have a conventional hexagonal head, a screw eye or a nut and washer. As the bolt or nut is tightened, the conical insert is drawn up within the metal sleeve to expand the anchor within its hole.

What to do

For fixings into masonry, make sure that the anchor will be inserted into a hole drilled in the centre of a brick, not into a mortar joint. Avoid positioning fixings in the top two courses of a brick wall, where the expansion forces could burst the brickwork. Drill a hole of the appropriate size in the brick and slide the anchor into place. Remove the bolt (or the nut and washer) and offer up whatever you are fixing. Tighten the bolt or nut with a spanner until you feel the anchor begin to bite within its hole, and give it a few more turns to complete the fixing.

You can also use expansion anchors to make fixings into concrete, although drilling the large diameter holes may be beyond the capacity of a DIY hammer drill. An alternative is to use rag bolts, which have one finned end designed to be embedded in the concrete before it has set. Once it is in place, a nut and washer are fitted to the exposed threaded end of the bolt to make the fixing. However, they can be difficult to place accurately.

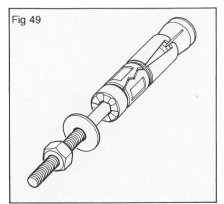

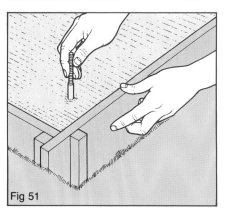

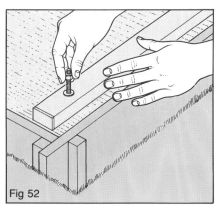

Fig 49 Expansion (masonry) anchors have a split metal sleeve, a cone-shaped plug and an integral bolt. The sleeve is forced apart by the plug as the bolt is tightened and draws it towards the outside of the hole.

Fig 50 Drill a hole to match the sleeve diameter, insert the anchor and remove the bolt or nut so you can offer up the fixture. Tighten the bolt or nut to make the fixing.

Fig 51 Use rag bolts to make fixings into fresh concrete. Push the finned tail of the bolt into the concrete.

Fig 52 Position what you are fixing over the threaded end of the bolt to hold it upright while the concrete hardens. Then complete the fixing by threading on the washer and nut, and tighten the nut down with a spanner.

Using Access Equipment

You will need a wide range of access equipment to carry out the various jobs in this book, especially as far as exterior and interior repairs and jobs such as plastering are concerned. For safety's sake, always use the correct equipment for the job, and never try to improvise; one fall could be your last.

The basic item for access to the upper parts of your house walls is a ladder. An aluminium extension ladder is ideal, since it is fairly light and easy to move around, reasonably sturdy to climb and very durable. If you are buying one, choose a size long enough for the top to reach to a point about 1m (3ft) above eaves-level with an overlap of four rungs – never less – between the extended sections. A triple ladder is often more useful than a double because you can separate the sections for access to different heights.

For low-level access – up to around the tops of ground-floor windows – you can either use steps or set up trestles to support scaffold boards in front of the area you are working on.

Ladder Safety

Falls from ladders are by far the biggest cause of DIY accidents, so do take care if you are using one for any maintenance or repair jobs. Set it up on level ground, with the foot of the ladder 1m (3ft) out from the wall for every 4m (13ft) of ladder height. Do not over-extend the ladder; you need an overlap of three or four rungs between sections with all extension ladders. Fit a stand-off near the top if you want the head of the ladder held away from the wall.

Wherever possible, secure the head of the ladder to a window-frame or to stout screw eyes fixed to the house wall just below eaves-level. Do not rely on down-pipes to provide a firm fixing. Weight or rope the foot too for extra safety.

Take care with low-level access equipment too – you can still have a bad fall from this. Always use proper scaffold boards rather than scrap timber or man-made boards, either set between two step-ladders or supported on hired adjustable trestles at the required height.

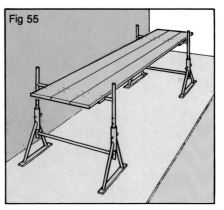

Fig 53 An extension ladder is a must for many outside repair and maintenance jobs. Add a stand-off to hold the top away from the house wall, a bracket for hanging tool carriers and a foot platform for extra comfort. Make sure the ladder is tied to the building at the top and is weighted or secured at the foot too.

Fig 54 For low-level jobs such as repointing brickwork, you can support a scaffold board on the rungs of a pair of stepladders.

Fig 55 Alternatively, hire trestles and boards to create a wider working platform – ideal for jobs such as plastering.

OUTDOOR JOBS

The great outdoors is definitely the place for the majority of your first building projects. Not only does it give you huge scope for improvement; it is the perfect environment for mastering the various skills you need to become a good bricklayer – or even a master mason – without having to worry unduly about the mess you are making. The garden is a very forgiving place, and you can retire to a warm tidy house at the end of a hard day's work without having to do all that tedious clearing-up and putting-away that always takes so long with indoor jobs!

The outdoor projects in this chapter are also extremely good value for money, for the simple reason that the materials you will be using are by and large relatively inexpensive. It is the labour content that pushes up the price; you can comfortably save yourself fifty per cent of a contractor's quote by doing the work yourself.

What You can Tackle

Armed with a few bricks and paving slabs and a bucket of mortar, the only limit to what you can achieve is your own imagination. You may want nothing more than a paved path down to the greenhouse or a dwarf wall round the patio, but if you are prepared to be a bit more adventurous you can create spacious patios, sweeping driveways, serpentine walls, ornamental arches, flights of steps to link different areas of the garden, even a garden pond. Here is a brief look at some of these ideas in more detail.

Patios, Paths and Drives The simplest jobs to tackle involve creating hard-surfaced areas on which to sit out, walk up and down or park your car. In some cases you can do this without even having to mix a bucket of mortar; all you need to do to create a simple patio or garden path is to bed paving slabs on a bed of sand, although this particular method is not recommended for a surface on which you intend to drive a car. But if you are prepared to be a bit more adventurous and hard-working you can lay interlocking paving blocks, crazy-paving or concrete in all sorts of shapes and sizes to create surfaces that complement your garden layout.

Walls and Steps The next stage is to begin to work in three dimensions, by building walls to enclose different areas of the garden, allowing you to create terraced areas and raised planters or to hide eyesores like the dustbin or the compost heap. Walls also provide excellent boundaries and ideal windbreaks, and can with a little ingenuity be designed to match the style of your house. Steps marry the skills of bricklayer and paver, and can allow you to make the most of even the most awkward sloping sites.

The materials you can use come in a wide variety of colours and textures, allowing you to build in formal brickwork or rustic stone as you wish.

Other Features You can add other masonry features such as garden arches, raised planters, even a garden pond, once you have mastered the basic skills of paving and bricklaying. There are also jobs such as casting concrete bases and supports that are not in themselves glamorous – or indeed even highly visible once completed – but which are a vital part of other projects such as erecting garden outbuildings, setting fence posts in concrete or even providing an in-ground socket so you can put up your rotary clothes line.

Fig 56 (*above*) Outdoor projects offer tremendous scope for individuality and can save you a lot of money into the bargain. All you need are some basic skills and the patience to plan and work methodically.

Planning Projects

Even the simplest outdoor projects need a little planning if the work is to proceed smoothly. For small-scale jobs such as building a straight section of wall or laying a path, you need little more than some measurements to enable you to estimate quantities, and a few pegs to mark out the site ready for excavation, laying foundations and so on. For larger-scale projects, however, there is no substitute for drawing up some proper detailed scale plans. Not only will this help you make major decisions about siting the various components of your scheme; it will also enable you to plan the various stages by which work will proceed, so avoiding silly mistakes like having to cart loads of walling blocks to the bottom of the garden across your newly-turfed lawn.

Where to Start

If you are good at visualizing things and have a clear idea of what you want in your garden, you can get the squared paper and pencil out and start drawing. But if you are less certain of your intentions, the golden rule is not to rush things. Study books and magazines for ideas, visit local garden centres to look at materials, stroll around public parks or visit stately homes to see how they make use of open spaces and formal features, even peer over fences at other people's gardens to see what they have been up to (it is polite to ask the owner's permission). Then start to give your ideas substance by thinking carefully about how you want to use your garden.

What will it actually be used for? The needs of a family with small children are obviously very different from those of a retired couple, although there is no reason why a garden should not change and grow older just like its owners. For example, areas initially set out as play space can be encroached upon as time goes by, and features can be added or removed at will. However, getting the major components right at the beginning is half the battle.

Think too about things like how the sun traverses the site, so you can make the most of direct sunlight falling on your

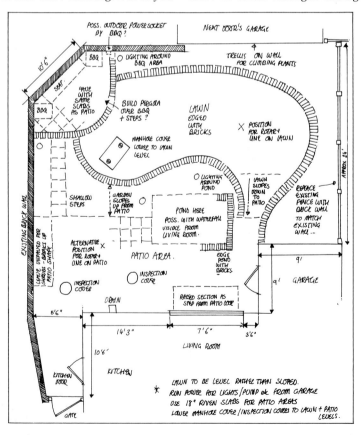

Fig 57 Your rough sketches can carry a lot of detail, both of what you intend to create and what is there already. Don't forget to include immovable obstacles such as drainage inspection chambers, or existing features such as mature trees that will remain.

proposed sitting-out area, your prize flowering shrubs or the site of your barbecue. If you are lucky enough to have a nice view, site your patio to take advantage of it; conversely, if next door is an eyesore, plan the layout to conceal it as far as possible and provide you with maximum privacy.

Next, give some thought to the garden's form. Do you want as big a lawn as you can get, or would you rather forego all that mowing and create a miniature woodland glade you can walk through at your leisure? If the site slopes, would you rather have a series of terraces than a continuous slope? Do you want a formal or informal look? What building materials will you use to achieve it? What about the boundaries? Do you want walls, fences, hedges or a mixed shrub border?

What other specific features would you like to include in your garden layout? A barbecue, a pond, a rock garden, a timber pergola are all possibilities, and then there is the question of outbuildings – a greenhouse perhaps, a shed for all your gardening equipment, even a small summer-house

could be part of your overall plan. So could the provision of underground services for things like outside lights and extra garden taps, which need thinking about well before work actually starts.

As you ask, and answer, these questions, you will find things beginning to gel and you can then start to make some rough sketches. As decisions begin to fall into place you can then start transferring your ideas to a detailed scale plan.

Drawing Plans

Work to a sensible scale – say 1:10 for a small garden, or 1:20 for a larger plot – and use metric measurements from the start for ease of reckoning. Make sure you know the sizes of the various materials you will be using – paving slabs, walling blocks and so on. Remember that your plans do not have to be miracles of draughtsmanship or works of art so long as they are accurate and make your intentions clear. They can always change as work proceeds and you realize that something won't work or you simply have a better idea!

Fig 59   Use a pickaxe and sledge hammer to demolish unwanted garden features.

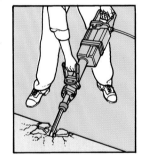

Fig 60   To break up large areas of concrete or crazy-paving, hire a power breaker to speed up the work and cut down on the effort

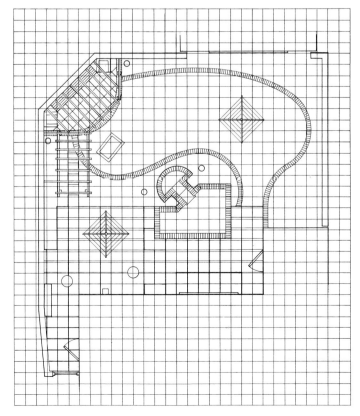

Fig 58   The final version is a plan to scale on squared paper that shows precisely where everything goes.

Laying Slabs on Sand

One of the simplest outdoor building projects is laying a patio or garden path on a bed of sand. It is also a good chance to get used to setting things out and handling materials because you can easily alter your plans or correct your mistakes as you go along. You can even take the slabs with you when you move house!

The sand bed serves two purposes. The first is to provide continuous support for the undersides of the slabs, and the second is to accommodate any slight unevenness in the underlying subsoil. Ideally you should use sharp concreting sand for the job, but soft sand will work just as well.

The one drawback with this method is that unless the sand is contained at the edges of the paved area, it tends to leach out as time goes by, causing the edge slabs to subside slightly. This is not a problem if the paved area is level with surrounding turf, for example, but you will need some form of edge restraint alongside slabs bordering flower beds to keep the sand bed in place. Refer to pages 38–9 for how to provide suitable edging.

What to do

Start by marking out the area to be paved with pegs and stringlines (see page 26), and then remove turf and topsoil from the site. Save both for re-use elsewhere in the garden. You need to excavate to a depth of about 100mm (4in) to allow room for a layer of sand 50–65mm (2–2½in) thick plus the slabs themselves. Make sure that the finished surface level of the patio will be at least 150mm (6in) below the level of the damp-proof course in the house wall, so that heavy rain cannot splash up above it off the patio surface; dig deeper if necessary to get this clearance.

If the subsoil is unstable or has been disturbed recently, you should dig down about another 75mm (3in) and then lay a bed of well-rammed hardcore over the area to stabilize it. Do not place it just yet, though.

The easiest way of getting an even thickness of sand is to divide the area to be covered into bays about 1.2m (4ft) wide with levelling boards held on edge with

What you need:
- paving slabs
- garden spade
- levelling boards and pegs
- mallet
- hardcore
- sharp sand
- wheelbarrow
- straight-edge
- brick bolster and club hammer or
- hired angle grinder and cutting discs
- soft broom

CHECK
- that the patio surface is at least 150mm (6in) below the house DPC.
- that it will finish level with any existing inspection chambers within the patio site.

TIP
To estimate how much sand you need for the job, measure the patio length and width in metres, multiply the two together to get the overall area and then divide by twenty to get the sand volume in cubic metres. Add ten per cent to allow for unevenness in the subsoil.

Fig 61

Fig 62

Fig 63

Fig 64

Fig 61   Excavate the site, then set out levelling boards across it at about 1.2m (4ft) intervals.

Fig 62   Tip out the sand and level it between the boards with a long timber straight-edge.

Fig 63   Remove the boards and fill the gaps with more sand.

Fig 64   Lay the first row of slabs next to the house, sliding each one into place off the edge of its neighbour.

Laying Slabs on Sand

pegs. Fit one against the house wall, another at the opposite edge of the site and at intervals in between. Check that they are level. Now add the hardcore if it is needed, followed by the sand, and use a timber straight-edge long enough to span your levelling boards to smooth the sand out. Kneel on a board as you do this, then lift out the boards and pegs and fill the gaps with a little more sand.

Lay the first row of slabs against the house wall, sliding each slab into place off the edge of the one next to it to avoid disturbing the sand bed unduly. Repeat the process for subsequent rows until you have completed paving the whole area.

If you have to cut any slabs in half, use a brick bolster and club hammer to score and then cut the slab. For awkward cuts – round inspection chambers, for example – it is worth hiring an angle grinder and some masonry cutting discs to do the job quickly and neatly.

Once all the slabs are in place, brush sand over the surface to fill the gaps and replace turf along the perimeter if necessary for a neat finish.

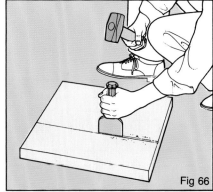

Fig 69 (*above*) Laying slabs on sand is a quick and simple way of laying patios and paths.

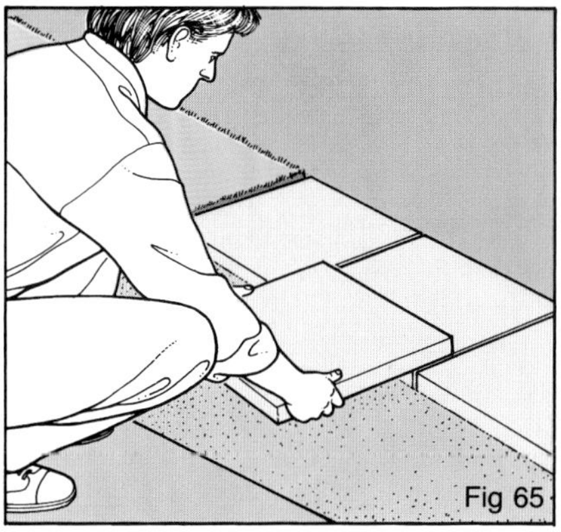

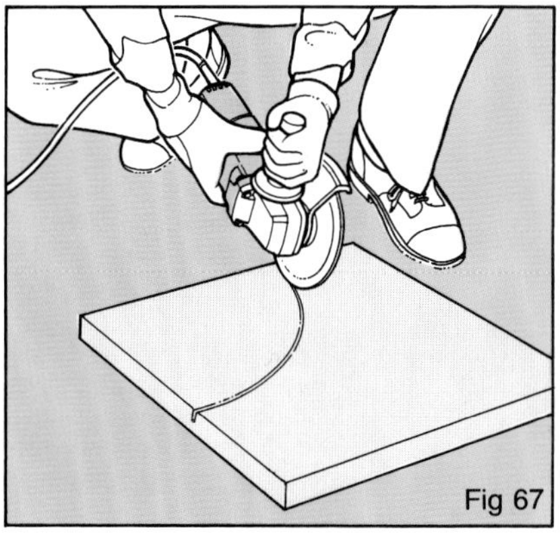

Fig 65 Lay subsequent rows in the same way, working off a wide board to spread your weight and avoid disturbing the sand bed.

Fig 66 Use a brick bolster and club hammer to score and break the slab if you need cut pieces.

Fig 67 For awkward cuts you will save time, effort and breakages by using an angle grinder.

Fig 68 Finish the job by brushing some sand into the joints between the slabs.

Bedding Slabs on Mortar

For areas that are likely to be heavily used, or where you are laying slabs over an existing old concrete surface, you will get better results by bedding the slabs on mortar. This does not have to be a continuous mortar bed for paths and patios (although you need one for drives); all you need is a dollop of mortar under each corner of the slab and one in the middle. However, you do need to set out your site carefully, to ensure that the paved area has a slight slope so rainwater will run off it. This does not matter with slabs laid on sand, since water can drain away between them. Aim for a slope of about 1 in 40 across the site – a 25mm (1in) fall for every 1m (3ft) of patio or path width.

With this method, you can place the mortar directly onto well-rammed subsoil. As with laying slabs on sand, check that the finished surface adjoining the house is at least 150mm (6in) below the level of the DPC, and take care to ensure that the slabs finish level with any manhole covers that fall within the paved area. You may have to raise these slightly if you are paving on an old concrete slab.

What to do

Start by marking out the site. Then excavate the turf and topsoil to a depth of about 25mm (1in) more than the thickness of the slabs you are laying. Check the fall across the site, and make any necessary adjustments. Then set up stringlines at each end of the site, parallel with the direction of the slope, to act as a guide while you lay the slabs. If you are laying the slabs over existing concrete which has no fall, you can introduce one by gradually increasing the thickness of the mortar pads as you work across the site.

Mix up some standard bricklaying mortar to a soft consistency, and place five fist-sized pats beneath the position of the first slab. Lower it carefully into place and tamp it down lightly with the heel of your bricklaying trowel. Use a spirit level to check that the slab is level in one direction and has the correct slope in the other. Then carry on laying slabs row by row, checking line and level as you proceed. Finish off by brushing dry mortar into the joints and watering it in.

What you need:
- paving slabs
- bricklaying mortar
- garden spade
- spirit level and straight-edge
- pegs and stringlines
- mallet
- bricklaying trowel
- brick bolster and club hammer *or* angle grinder and cutting discs
- soft broom
- watering-can

CHECK
- that the fall across the site remains constant by referring to your stringline as you lay the slabs. Lift any slabs that are high or low, removing or adding mortar as necessary.
- that dry mortar is brushed off the paved surface before you water it, or staining will result.

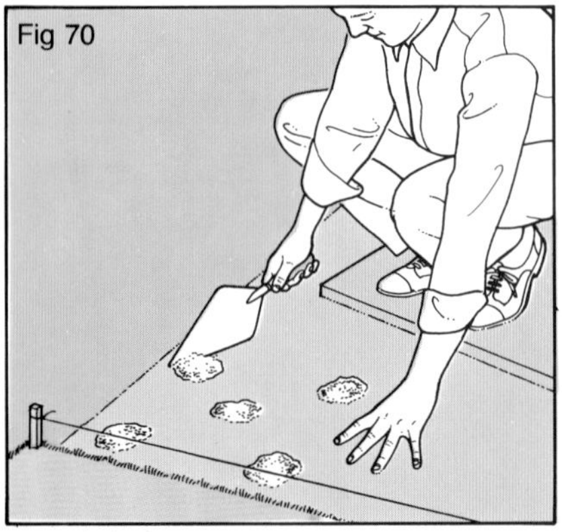

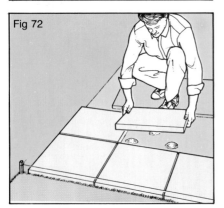

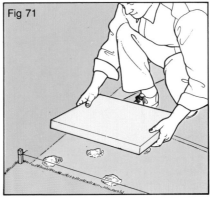

Fig 70 After excavating and levelling the site, place five pats of mortar on the subsoil beneath the position of the first slab.

Fig 71 Lower the first slab into place, tamp it down and check it with your spirit level.

Fig 72 Carry on laying the rest of the slabs in the same way, checking for level along the site and slope across it as you proceed.

Fig 73 When all the slabs have been laid, brush a dry 1:5 cement:sand mix into all the joints and sprinkle water over the paved surface to help make the mortar set.

Laying Crazy Paving

Crazy paving is a curious name for a way of creating paved surfaces using interlocking pieces of irregularly-shaped stone – usually broken paving slabs, although natural stone can be used instead. It may have got its name from the archaic meaning of crazy as broken down or dilapidated (reflecting the look of much old paving of this type), or simply by analogy with the crazing of the glaze on old pottery, which it resembles on a somewhat larger scale!

The main advantage of crazy paving is that the raw material is cheap. Its big drawback is that laying it takes longer to lay than any other type of paving, mainly because of the laborious pointing required to fill the gaps between the pieces. This makes it an ideal do-it-yourself project, where labour costs are effectively free.

The best source of crazy paving is local demolition contractors, and also some local authorities who use it as a way of recycling municipal paving materials. When ordering, specify whether you want pieces of a uniform colour and texture, or whether you prefer mixed colours.

What to do

Crazy paving must be laid on a continuous mortar bed, either on a sub-base of well-rammed subsoil or hardcore, or over an existing concrete surface. Estimate mortar quantities by allowing for a mortar bed 50mm (2in) thick, plus half as much mortar again for the pointing.

Mark out and excavate the site (if necessary), levelling compact subsoil or ramming hardcore into unstable soil and then blinding it with a layer of sand.

When the paving arrives, sort it into three separate piles. One should contain medium to large pieces with one straight edge, the second other medium-sized pieces and the third the smaller leftovers. Use stones from the first pile to form the edges of the paved area; pieces from the second pile will fill in most of the centre of the area, while the smaller pieces from the third pile fill the gaps.

Aim to lay all the stones in their mortar bed first, then point the joints after about forty-eight hours.

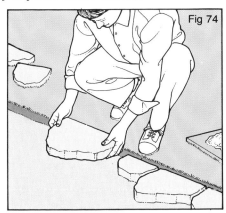

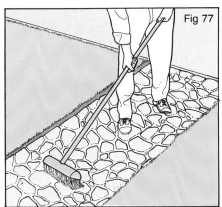

Fig 74 Start by placing large slabs with one straight edge along the perimeter of the area you are paving. Use a continuous mortar bed over subsoil or rammed hardcore, mortar dabs on a concrete base.

Fig 75 Infill the centre area with large, then small pieces. Check that all sit level with their neighbours.

Fig 76 When the bedding mortar has hardened, point between the stones with a fairly dry mortar mix.

Fig 77 When the pointing is dry, brush off any excess mortar.

Laying Block Paving

Block paving has been the big success story of recent years as far as hard surfaces are concerned. It first began to appear in public open spaces such as shopping precincts, and then spread to the new housing market before being taken up by the do-it-yourselfer. The beauty of it is that the blocks are relatively small and manageable, unlike paving slabs, and that they are dry-laid – no mortar and no pointing. Best of all, unlike other dry-laid paving they can withstand the weight of motor vehicles, so they can be used for all hard surfaces around the house and garden. Add to that the fact that they come in almost as many colours and textures as bricks, and you have virtually the ideal paving material.

The one drawback is that since the blocks are laid on a sand bed, you need to provide some form of edging all round any paved areas, even where they abut turf, to prevent the sand from leaching away and allowing the blocks to subside.

However, this is a small price to pay for choosing an otherwise excellent general-purpose paving material.

What to do

Mark out and excavate the area as for any paving project. The depth you need is 50mm (2in) plus the thickness of the blocks themselves – typically around 65mm (2½in). Subsoil should be undisturbed below paths and patios; for drives you need to excavate a further 100mm (4in) on sandy soils and deeper still on clay to prevent subsidence, and then put down a layer of well-compacted aggregate.

Next, position the edge restraints all round the area to be paved. You can use proprietary path kerbstones, a row of the paving blocks laid end to end or side by side, both bedded in concrete, or even preservative-treated timber planks secured with stout pegs. As you position the edging, check that each unit is accurately aligned and that opposite sides of square or rectangular sites are truly parallel to each other. If you are setting the edging in concrete, keep this at least 65mm (2½in) below the top surface of the edging to allow the sand bed to be laid right up to it.

What you need:
- paving blocks
- spade and pickaxe
- aggregate or hardcore
- edge restraints
- fine concrete
- bricklaying trowel
- spot board
- spirit level and straight-edge
- sharp sand
- levelling batten
- brick bolster and club hammer *or*
- hired hydraulic block splitter
- soft broom
- hired plate compactor

CHECK
- that blocks are laid tightly butted against their neighbours, and are level with them.

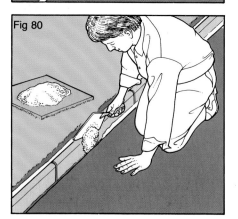

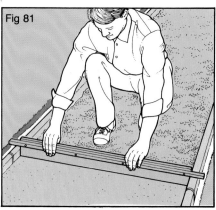

Fig 78   Start by positioning the edge restraints along each side of the area to be paved.

Fig 79   Check that the edging units are correctly aligned and that opposite edges are level with each other and also parallel where necessary.

Fig 80   Secure kerb stones with a haunching of fine concrete on each side. Keep the concrete low enough to allow the sand bed to be laid right up to the edging.

Fig 81   Use a notched batten to spread and level the sand bed so it is a little more than the thickness of the blocks below the top of the edging.

Laying Block Paving

When the concrete has hardened, lay the sand bed across the site and use a notched batten to level it to the required depth below the edge restraints. Then you can start laying the first blocks.

If you are working with a pattern that runs parallel to the edging, simply build it up by working across the area row by row. Set stringlines to guide you if you are working with diagonal patterns.

For a light-duty area, simply tamp each block down firmly into the sand bed with a club hammer. For areas that will carry heavier traffic (including driveways) you will have to use a hired plate compactor instead to vibrate and bed the blocks fully into the sand.

When all the whole blocks are laid, split and fit any cut pieces that are needed to complete the pattern. It is best to hire a hydraulic block splitter if you have a lot of cutting to do; otherwise use a brick bolster and club hammer. Finish off by brushing sand into the joints, and run the plate compactor over the finished surface to vibrate sand deep into them.

Fig 86 (*above*) Block paving is the only loose-laid type that can safely support a car's weight.

Fig 82

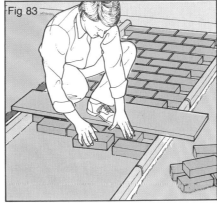

Fig 83

Fig 82 Start laying the blocks against the edging at one side. Use a stringline as a guide with diagonal patterns.

Fig 83 Lay all the whole blocks first, then split and fit smaller pieces as necessary to fill in any gaps.

Fig 84

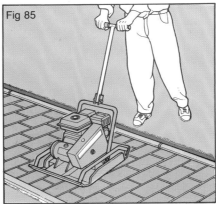

Fig 85

Fig 84 On light-duty areas, tamp blocks down with a club hammer and a block of wood.

Fig 85 On heavy-duty areas, run a hired plate compactor over the surface to ensure that the blocks are fully bedded into place. Finish the job by brushing sand into the joints and then run the compactor over the surface again.

Surfacing a Drive

You can of course lay a driveway with any of the paving materials already discussed, but there is an alternative to masonry: it is what most people call Tarmac (short for tar macadam, and named after its inventor) or asphalt. This is crushed stone or fine aggregate coated with a bituminous binder, and is spread out and rolled flat to create a smooth, seamless and very hardwearing surface.

For large areas, it is best to order hot asphalt from a local supplier, but for small areas such as a garden path you can buy cold asphalt in bags. Both types are available in black, with red (and green cold asphalt) being less common. Whichever you use, you can also buy coloured chippings to roll into the finished surface which give an attractive speckled appearance.

The main advantages of asphalt as a drive material are that it is relatively quick to lay, needs minimal maintenance and is easy to repair. Its disadvantages are that it can become tacky in hot weather or if petrol or oil are spilt on it, and it dents under point loads.

What to do

If you plan to lay hot asphalt, tell your supplier the area to be covered and what the subsurface is, in case an initial priming coat of bitumen emulsion is needed. Tell him also that you will be unable to keep the load hot, so he can include special additives to keep the mix pliable for longer. Hire an eight-pronged fork, a wide rake and a punner, plus a heavy-duty wheelbarrow, a large tarpaulin, a plate compactor and a motorized hand roller if you intend to lay the material yourself.

You can lay hot asphalt over concrete and mortared paving, or over well-rammed subsoil topped with hardcore. Sweep hard surfaces, fill potholes and apply weedkiller before you start the actual laying process. You will need edge restraints as for block paving (see pages 38–9) if the drive abuts a surface at a lower level, to stop the edges from crumbling.

Hot asphalt is laid to a thickness of about 38mm (1½in), so you will need to raise the top frames holding inspection

Fig 87

Fig 88

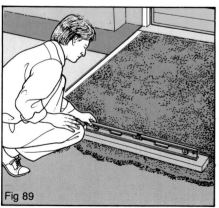

Fig 89

Fig 90

Fig 87   Load your barrow from the heap using an eight-pronged fork. Keep the heap covered with a tarpaulin to stop it cooling too quickly.

Fig 88   Fill potholes first using your punner. Then rake the material out to a thickness of about 50mm (2in), using the back of the rake to break up lumps.

Fig 89   Check the level regularly with a spirit level and straight-edge.

Fig 90   Run over the finished surface with the plate compactor first, then scatter chippings and roll it with the heavy roller.

Surfacing a Drive

chamber covers by this amount, and you may also have to trim outward-opening garage doors (*before* laying the asphalt). Block nearby gullies with rags to stop loose asphalt getting into the drains.

When the load arrives, keep the heap covered with the tarpaulin. Dump barrow-loads of material at intervals across the drive, rake it out to about 50mm (2in) thick and use the back of the rake to level it. Check the level as you work to ensure a slight fall for drainage purposes. Then compact it with the plate compactor, scatter chippings over the surface and run over it with the heavy roller.

Cold asphalt is much easier to lay over existing concrete or macadam, but is not very successful on a hardcore base. Coat the area with bituminous emulsion, then simply tip the material out of its bags, rake it out to a thickness of about 25mm (1in) and then compact it with a garden roller to a finished thickness of about 19mm (¾in).

If you just want to give old concrete or macadam a facelift, simply top-dress it with bitumen emulsion and fine chippings.

Fig 95 (*above*) Decorative coloured chippings give an attractive speckled appearance to the drive surface.

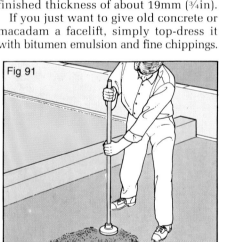

Fig 91

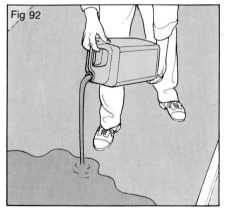

Fig 92

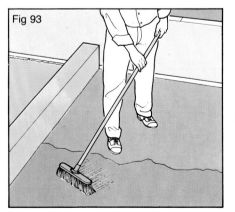

Fig 93

Fig 94

Fig 91 With cold asphalt, fill hollows first using your punner.

Fig 92 Pour out the bituminous emulsion.

Fig 93 Use an old broom to spread it evenly over the area. Let it dry for about twenty minutes.

Fig 94 Rake out the asphalt to a thickness of about 25mm (1in), then roll it with your garden roller. Keep the roller wet using a watering-can to stop the material sticking to it. Scatter chippings and roll again.

Building Free-Standing Walls

If you intend to build any free-standing brickwork, especially runs of walling, it is essential to pay careful attention to its design to ensure that it performs as you expect it to. In particular, this means building it off suitable foundations, and also including piers to give it strength.

All walls need foundations in the form of a concrete strip poured into a trench. For walls up to about 750mm (2ft 6in) high, the width of this strip should be twice the width of the wall being built on it, and should be a minimum of 150mm (6in) thick. For walls higher than 750mm (2ft 6in), the strip width should be three times the width of the wall. In both cases its top surface should be 150–225mm (6–9in), or two to three brick courses below ground level to allow soil or turf to be replaced right up to the foot of the wall.

On sloping ground you will have to lay foundations in stepped form. Cast each step in turn from the bottom upwards, with a horizontal overlap equal to one brick or block length and the step height equal to the brick or block height.

Building End Piers

You need piers at the ends of any wall, even a low one where a one-brick square pier should be included (see Fig 96 and page 24). In exposed locations, it is better to build piers one-and-a-half bricks square on the centre line of the wall to provide extra stability whichever way the wind blows, and piers of this type can also be reinforced by filling the hollow centre with concrete or by building them round a steel reinforcing rod set in the foundations. For extra strength, tie the wall to the piers with strips of expanded metal mesh set in the bedding mortar every two or three courses.

Fig 96   All walls need concrete footing at least twice as wide as the brickwork and at least 150mm (6in) thick.

Fig 97   On slopes, cast the strip in steps with overlaps as long and high as the walling unit being used.

Fig 98   How to build up centred one-and-a-half brick end piers with ties bonding the wall.

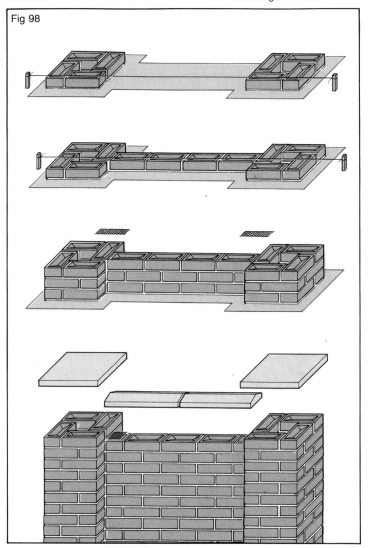

Fig 98

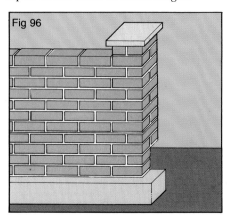

Fig 96

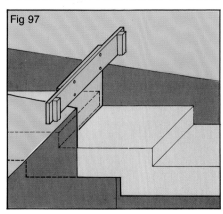

Fig 97

Building Free-Standing Walls

Intermediate Piers

If the wall is being built in stretcher bond 100mm (4in) thick, and the height exceeds 450mm (18in), you should also incorporate intermediate piers at 3m (10ft) intervals. Again, one-brick piers can be used, but centred one-and-a-half brick piers offer extra stability in exposed locations. The maximum height you should build in 100mm (4in) thick brickwork with one-brick piers is 675mm (27in); for 215mm (9in) thick brickwork you can build to a height of 1.35m (4ft 6in) without inter-mediate piers and to 1.8m (6ft) with them. You should seek expert advice on the construction of any wall higher than this.

Movement Joints

On straight runs of walling over 6m (20ft) long you should incorporate movement joints to prevent settlement or temperature changes causing buckling of the wall struc-ture. Form them where the wall meets a pier by including a 300 × 25mm (12 × 1in) strip of greased galvanized metal with-in the mortar courses.

Fig 103 (*above*) Free-standing walls provide the perfect boundary for any property frontage.

> **TIP**
> Do not build up the wall height by more than 1m (3ft 3in) in a day, to avoid excess weight on the fresh mortar.

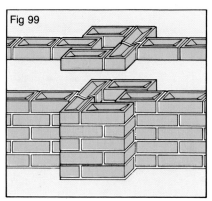

Fig 99

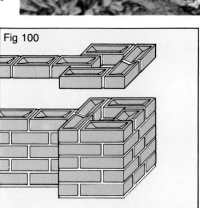

Fig 100

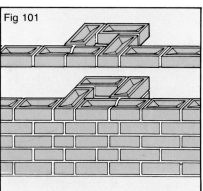

Fig 101

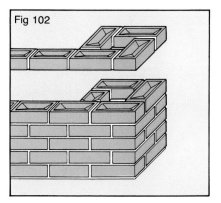

Fig 102

Fig 99   How to build up centred one-and-a-half brick intermediate piers bonded into the wall.

Fig 100   How to build up centred one-and-a-half brick end piers bonded into the wall.

Fig 101   How to build up offset one-and-a-half brick intermediate piers bonded into the wall.

Fig 102   How to build up offset one-and-a-half brick end piers bonded into the wall.

Building Screen Block Walls

Pierced screen walling blocks offer an attractive alternative to solid masonry blocks, especially where an open screen effect is wanted. They can be used on their own or as decorative infill panels in walls built of other materials.

The big difference between these blocks and other walling units is that since they are square there is no horizontal bonding between the units; they are simply stacked up. This stack bonding is naturally weak, and so piers (built using special grooved pier blocks) must be incorporated at a maximum of 3m (10ft) intervals, and horizontal reinforcement must be used on walls more than two blocks high to tie wall and pier together after every two courses of blocks (which coincides in height with three pier blocks).

In addition, reinforcement should be incorporated within the piers on walls more than two blocks high, in the form of 16mm (⅝in) diameter steel rods or 50mm (2in) square angle irons, set in the foundation strip at each pier position. The piers are then filled with mortar.

Fig 104 (*above*) Pierced block walls are ideal for screens that allow light and air to pass freely.

What you need:
- wall and pier blocks
- coping and pier caps
- reinforcing rods
- mortar – *see* Tip
- bricklaying trowel
- spirit level
- pointing trowel

TIP
Use white Portland cement and white sand to make a mortar that will match the block colour. Alternatively, paint the blocks with masonry paint.

Fig 105 Screen walling systems include square blocks, grooved pier blocks for end and intermediate piers and corners, coping stones and matching pier caps.

Building Screen Block Walls

What to do

Start by casting your foundation strip to finish at ground level, with reinforcing rods built in at all pier positions. Then build up the first pier to a height of 600mm (24in) (three blocks), and position the first walling block in the pier's side groove after buttering mortar onto its edge. Add a second block on top of the first and check that it sits level with the top of the pier.

Complete the rest of the first course and start building up the next pier before adding the second course.

If you are building higher than two courses, use expanded metal mesh to bond the top of the second course to the pier at each end. Then add two more courses and leave the mortar to harden overnight. Complete the remaining courses the next day and add coping stones and pier caps to complete the wall.

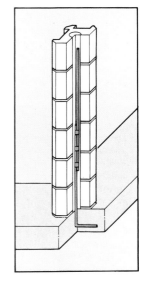

Fig 112   On walls over 600mm (2ft) high, piers must be reinforced with steel rod or angle iron set in the foundations.

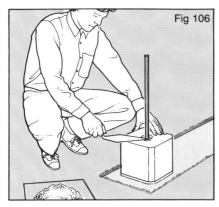

Fig 106

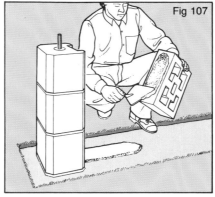

Fig 107

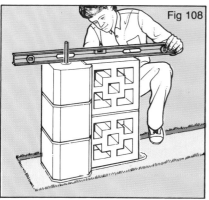

Fig 108

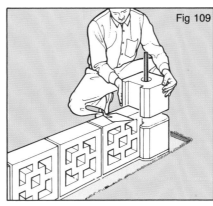

Fig 109

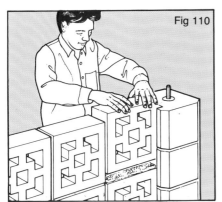

Fig 110

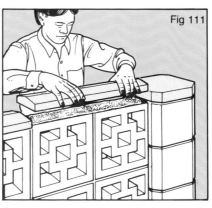

Fig 111

Fig 106   Set the first pier block over the reinforcing rod on a mortar bed, with the groove facing the wall.

Fig 107   Build up the pier three blocks high, then mortar the side of the first block and set it in place on a mortar bed so it fits into the groove in the pier

Fig 108   Add a second block on top of the first and check that it is level with the pier.

Fig 109   Complete the first course, then build up the other pier.

Fig 110   Complete the second course. Tie both piers to the wall with expanded metal mesh set in the bedding mortar if the wall is going higher.

Fig 111   Finally, add copings and pier caps.

Building Steps in a Bank

If you have a steeply sloping garden, you will need steps to enable you to get from one level to the next easily and safely, and one of the simplest ways of constructing them is to use the bank itself to form the steps' foundations.

The steps should conform to the usual rules for any flight of steps or stairs, indoors or outside, with treads and risers of the same size throughout the flight. Make the treads at least 300mm (12in) from front to back and a minimum of 600mm (2ft) wide, or 1.5m (5ft) wide if you expect people to want to pass each other on the flight. Riser height will be governed by the size of the bricks or blocks you use to build them, but generally a height of 100–150mm (4–6in) is ideal – two courses of brick or standard walling block. Treads should overhang risers at the front by at least 25mm (1in), and should slope very slightly towards the front so water cannot stand or freeze on them and make them dangerous to use. Lastly, you should incorporate a wide landing halfway up on flights of more than ten steps.

What to do

Start by selecting your materials. Bricks or walling blocks that match other masonry in the garden are best for the risers; make sure bricks are frost-proof. For the treads paving slabs are the easiest to lay.

At the site of the steps, drive a peg in at the top of the bank and use a horizontal stringline and a vertical garden cane to measure the height of the bank. Divide the height of each step (riser plus tread thickness) into this to work out how many steps you will need. Then measure the length of the stringline to determine the bank length; divide this by the number of steps and check that the tread depth will be at least 300mm (12in). On steep banks you may need higher risers and fewer steps to achieve this minimum depth.

Next, set two stringlines between pegs to mark the line of the flight, and set more stringlines across the bank to mark the nosing positions. Start digging at the top of the bank, cutting out even step shapes with the risers in line with the cross strings.

What you need:
- paving slabs for treads
- blocks for risers
- pegs and stringlines
- tape-measure
- garden spade
- concrete
- bricklaying trowel
- mortar
- spirit level
- hardcore and punner
- broom

CHECK
- that treads are at least 300mm (12in) deep and 600mm (2ft) wide, project over the risers by 25mm (1in) and slope slightly to the front.
- that risers are 100–150mm (4–6in) high if the shape of the slope allows.

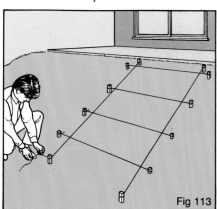

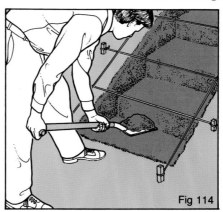

Fig 113   Set stringlines between pegs to mark the sides of the flight and the positions of the tread nosings.

Fig 114   Cut step shapes into the bank, starting at the top and working downwards so you do not break down the step edges by treading on them.

Fig 115   Cast a concrete strip foundation at the foot of the flight and then bed the blocks that will form the first riser on it.

Fig 116   Infill behind the first riser with hardcore, and ram more in on top of the next step to form a firm base for the tread.

Building Steps in a Bank

Remove these temporarily if they get in the way as you are digging.

Cast a foundation strip for the lowest riser, then build it up on a mortar bed. Carefully back-fill behind it with fine hardcore, then bed the first tread slabs in position on mortar. Check that they are level across the flight and have a slight slope towards the nosing.

Build up the second riser on the back edge of the first tread, then fill behind it with hardcore as before and place the next set of slabs on their mortar bed. Carry on building the flight up in this way until you reach the top of the bank. Finish off by pointing the joints on the risers neatly, and brush dry mortar into the joints between the tread slabs. Allow the mortar to harden for forty-eight hours before you start using the flight.

If the cut sides of the bank show signs of crumbling, build up dwarf side walls at each side of the treads to contain the soil. If the flight is steeper than about 30°, it is a good idea to add a post-and-rail handrail at one side.

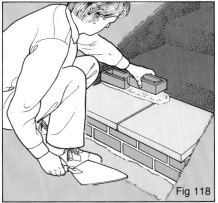

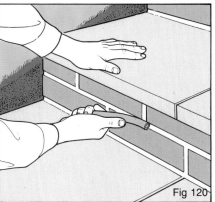

Fig 121 (*above*)   Steps are essential if you have a sloping garden.

Fig 117   Bed the first tread slabs in place on mortar. Check that they are level across their width and slope slightly towards the front.

Fig 118   Build the second riser up on the rear edge of the first tread slabs.

Fig 119   Add the second tread in the same way as the first, and continue the sequence until you reach the top of the flight.

Fig 120   Finish the job by neatening up the pointing on the risers, and brush dry mortar into the joints between the slab treads.

Building Free-Standing Steps

If you have a raised patio, or a series of terraces formed with earth-retaining walls in a sloping garden, you may prefer to have flights of steps built out from the patio or garden walls to provide access between the different levels. Such steps are more complex to construct than steps built into a bank, since you have to build up the shape of the flight from scratch instead of using the bank as a foundation.

It makes sense to plan the shape and size of the flight around the materials you intend to use to build it – ideally the same as those used for patio surfaces or for retaining walls, since the steps will form an extension to the existing masonry. The bricks or blocks will form the support for the flight, so a dry run to see how the bonding should be arranged is the best starting point. You can build up the flight course by course, and also lay the treads in place, with the aim of minimizing the amount of cutting involved and also of satisfying the basic provisions for good step design – see Check and page 46 for more details.

What to do

Once you have planned the construction in the dry, you can start the actual building work. A free-standing flight of steps needs sound foundations not only for the perimeter brickwork, but also for the bricks or blocks forming the intermediate risers. The simplest solution is to cast a concrete slab foundation about 100mm (4in) larger all round than the plan size of the flight, and 100mm (4in) thick – see page 26.

Once this has hardened, set out stringlines to help you position the first course of bricks accurately, and lay it on a mortar bed using stretcher bond. Follow this with the second course, just as if you were building up a dwarf wall.

To link the flight physically to the wall against which it is being built, you should bond the end bricks forming the sides of the flight into half-brick-sized recesses chopped in the face of the wall on every alternate course – a process known as toothing-in. This helps to prevent the risk of the steps starting to move away from

What you need:
- paving slabs for treads
- bricks or blocks for the flight walls
- pegs and stringlines
- tape-measure
- garden spade
- concrete
- hardcore and punner
- bricklaying trowel
- mortar
- spirit level
- pointing trowel

CHECK
- that treads are at least 300mm (12in) deep and 600mm (2ft) wide, project over the risers by 25mm (1in) and slope slightly to the front.
- that risers are 100–150mm (4–6in) high.

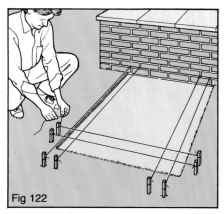

Fig 122

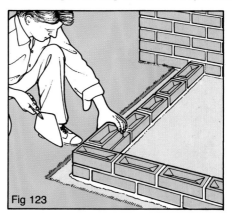

Fig 123

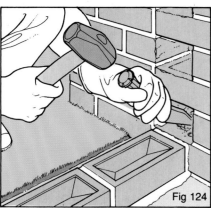

Fig 124

Fig 125

Fig 122   Lay a concrete foundation slab about 100mm (4in) bigger than the flight plan size, then set out stringlines to show the position of the first course of brickwork.

Fig 123   Lay the first course of bricks in a U-shape, then add the second course to form the first riser.

Fig 124   Complete the second course, toothing the last bricks into recesses chopped in the wall against which the flight is being built.

Fig 125   Build up the support brickwork for the second riser, and back-fill between it and the first riser with hardcore.

Building Free-Standing Steps

the wall through differential settlement.

When the flight is built up to the level of the first step, add two courses of brickwork inside the perimeter wall to support the second riser, and back-fill between it and the first riser with hardcore topped with concrete.

Next, add courses three and four to form the second riser and to raise the sides of the flight. Again tooth the fourth course into the back wall. Then build up the support brickwork for the third riser from the base slab, and infill between it and the second riser as before. Continue in this way until you reach the top of the flight and the brickwork is complete. Depending on the design of your flight, the topmost step may be level with the patio surface, or may finish against the wall so that the final step up is onto the patio.

Complete the flight by bedding the treads in place on a mortar bed. Check that each one overlaps the riser beneath it by about 25mm (1in), and that the treads are level across the flight and have a slight slope towards the front edge for drainage.

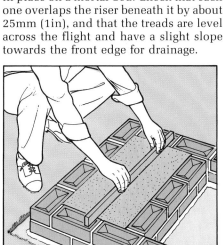

Fig 126

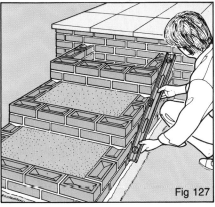

Fig 127

Fig 128

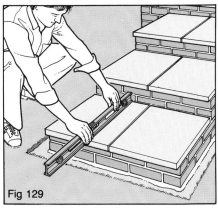

Fig 129

Fig 130 (*above*) Free-standing steps make a graceful addition to any terraced garden layout.

Fig 126 Top the hardcore with a layer of fine concrete.

Fig 127 Build up the flight by adding two more courses of bricks for each riser, and back-fill behind each one as before. Check that the brickwork is built truly vertical.

Fig 128 Lay the tread slabs on a mortar bed, checking that they overlap the riser below by about 25mm (1in).

Fig 129 Make sure the treads are level across the flight and slope slightly down towards the front edge.

Building Retaining Walls

If you have a sloping garden and you want to create level areas – to form a patio, create a flat lawn area or make some manageable flower beds, for example – then you will have to create a terrace effect by building retaining walls across the slope. Unlike ordinary free-standing walls, these must be built strongly enough to withstand the downward forces from the earth retained behind the wall, and these forces can be considerable, especially when the soil is waterlogged. There are several ways of achieving the necessary strength in any retaining walls you build, using bricks, garden walling blocks, hollow concrete blocks or natural stone.

On a gentle slope the precise siting of a single retaining wall is not particularly critical, and you can build it wherever it works best in forming an upper and lower level. However, on more steeply-sloping sites you will be trying to create a series of steps down the slope, and more care will be needed to get the best results. As a general guide, go for frequent shallow steps rather than one or two high ones.

What to do

Start by deciding on the positions for the walls you want to build, and mark these on the site with pegs and stringlines. If you are building a series of walls, start with the one furthest from your access point for materials, so you do not have to hump barrows from one terrace to the next.

Then excavate a trench for the concrete foundation, ram in a 100–150mm (4–6in) layer of hardcore and top this with a layer of concrete at least 150mm (6in) thick. If you have clay soil, increase the thickness to 300mm (12in) to ensure that the wall remains stable if the subsoil shrinks or heaves with changes in the weather.

Build in reinforcing rods if they are needed as you lay the concrete, propping them so they remain vertical while the concrete sets. Then you can start building.

If you are using bricks, make sure they are frost-proof. Remember that you need expensive facing bricks only for the outer face of the wall. For low walls up to about 600mm (2ft) high, build the wall without

What you need:

- pegs and stringlines
- spade
- hardcore
- concrete
- reinforcing rods
- bricks, blocks or natural stone
- bricklaying trowel
- mortar
- brick bolster and club hammer
- pipe for weepholes
- polythene sheeting or liquid waterproofer

Fig 131

Fig 132

Fig 133

Fig 134

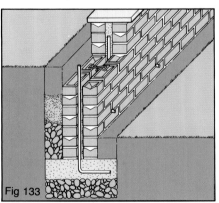

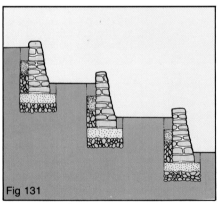

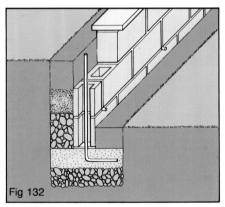

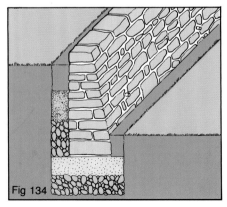

Fig 131 Use a series of earth-retaining walls in sloping gardens to create level terraces.

Fig 132 If you are using hollow concrete walling blocks and reinforcing rods, set the blocks over the rods and fill the block cavities with concrete.

Fig 133 For brick walls over about 600mm (2ft) high, build two parallel 'skins' of brickwork so they sandwich the reinforcing rods, and fill the cavity in between with concrete as the wall rises.

Fig 134 With drystone walls, construct the wall so it slopes backwards slightly into the slope.

Fig 135 (*right, top*) Foundations should be a minimum of 150mm (6in) thick and at least twice as wide as the wall, and should be laid over 150mm (6in) of well-rammed hardcore.

Fig 136 (*far right, top*) Where low walls adjoin a paved area, leave a drainage channel along the face of the wall so water does not stand on the paving.

Building Retaining Walls

reinforcement in 215mm (9in) thick brick-work, using one of the bonds illustrated on page 25. For walls higher than this – up to around 1.2m (4ft) high – build the wall as two parallel structures about 50mm (2in) apart in stretcher bond, so the re-inforcing rods are sandwiched between the two skins of brickwork. Tie the two together with cavity wall ties set in the bedding mortar every 450mm (18in) hori-zontally and vertically, and fill the cavity with fine concrete. Get professional advice on constructing retaining walls more than about 1.2m (4ft) high.

Use a similar technique with hollow wall blocks, placing them over the reinforcing rods and filling all the block cavities with fine concrete as the wall rises.

A drystone wall gets its strength from being constructed leaning backwards slightly into the slope and by the combin-ation of long 'through' stones running the whole width of the wall and smaller inter-locking edging stones.

Include weepholes at intervals along the wall to allow water behind it to escape, and back-fill behind the wall with gravel.

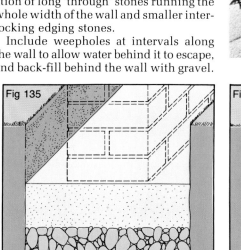

Fig 135

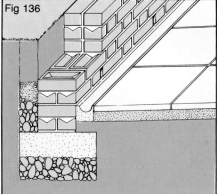

Fig 136

Fig 139 (*above*) Retaining walls are the perfect way of creating raised flower beds round a patio.

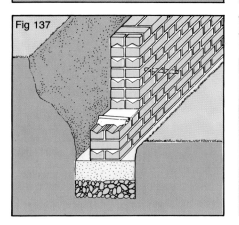

Fig 137

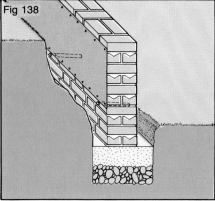

Fig 138

Fig 137 Build short lengths of copper or plastic pipe into the wall near the base to act as weepholes and allow groundwater to drain through from behind the wall.

Fig 138 Line the back face of the wall with polythene sheeting or liquid damp-proofer before back-filling with gravel or other granular material. This will help to prevent moisture from the retained soil soaking through the brickwork and causing unsightly efflorescence on the front face.

Building Brick Arches

One of the most elegant ways of forming openings in high walls or framing a gateway is to build a round-topped arch. The structure works by transferring the load of the arch (and any brickwork above it) downwards into the wall or piers supporting it. In a wall, the surrounding masonry is strong enough to carry and dissipate this load; if the arch is built as a free-standing structure the piers at the side must be at least 215mm (8½in) – one brick – square, and must have their own concrete pad foundations 450mm (18in) square and 150mm (6in) deep.

The secret of successful arch building lies in the cunning use of a curved former to support the arch brickwork as the mortar sets. The commonest arch profile in use is semi-circular, but there is no reason why you should not adapt the technique shown here to build what is called a segmental arch – one where the curved section is a smaller part of the perimeter of a circle.

You can build your arch using either bricks or small garden walling blocks to match other masonry nearby.

What to do

Start by deciding on the site of your arch and the size and profile you want to build. Then cast the pier or wall foundations, and build up the brickwork to the height at which the curve of the arch will begin – known as the springing point. This is commonly about 1.5m (5ft) above ground level for a semi-circular garden arch, allowing ample headroom even for six-footers when the arch itself is completed.

Next, cut two plywood or chipboard shapes to match the profile of your arch. Link the straight edges with a rectangle of wood slightly narrower than the thickness of the arch masonry, and add spacers cut from timber offcuts to brace the former and create a rigid structure.

Support the former between the piers using timber props so its base sits level with the tops of the piers. Then use a brick to mark the positions of the bricks and 10mm (⅜in) mortar gaps on one face of the former, so you can see how many whole bricks will be needed for the first

What you need:
- tape-measure
- spade
- concrete and hardcore for foundations
- bricks or blocks
- mortar
- bricklaying trowel
- spirit level
- plywood or chipboard plus timber offcuts and nails for former
- jigsaw and hammer
- timber props
- pointing trowel

TIP
If you have difficulty fitting a brick as the keystone, use cut pieces of slate, plain clay roof tiles or quarry tiles instead.

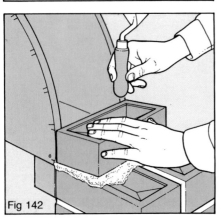

Fig 140

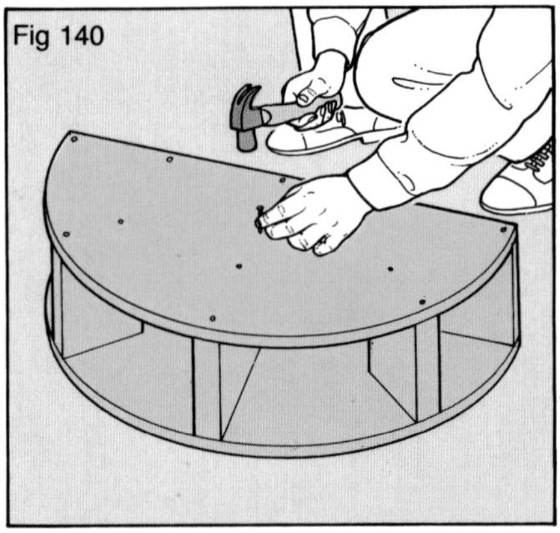

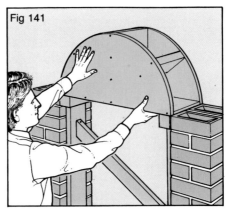

Fig 141

Fig 143

Fig 142

Fig 140 Build up the brickwork at each side of the arch until you reach the springing point. Then nail together a board and softwood former to match the arch width.

Fig 141 Set the former in place level with the piers, and support it with timber props. Mark the positions of bricks in the inner ring on one face of the former.

Fig 142 Start building the inner ring, placing the first brick against the former in line with the pencil marks.

Fig 143 Build up the ring, working from each pier up towards the top.

Building Brick Arches

(inner) ring of the arch. Work up from the piers at each side towards the top of the arch. If the space for the last brick at the top, the keystone, is a little too wide or narrow, try altering the thickness of the mortar joints slightly. Otherwise you can use pieces of slate, quarry tile or plain roofing tiles to form the keystone.

When you are happy with the arrangement, start building up the inner ring, working from opposite sides of the arch towards the top. When you reach it, butter mortar onto both sides of the keystone and tamp it into place. Check that the face of the arch is flat and truly vertical.

Next, spread a thin mortar bed on top of the inner ring and start building up the outer ring in the same way. Because this has a larger radius than the inner ring, the brick courses will be out of alignment and you will need more bricks – typically twenty-four for the outer ring and seventeen for the inner ring on a semi-circular arch 915mm (3ft) wide. Finish off by neatening the pointing. Leave the former in place for forty-eight hours, then remove it and point the underside of the arch to complete the job.

Fig 148 (*above*) A brick arch is the perfect frame for a garden gate. Here plain clay roof tiles form a decorative keystone.

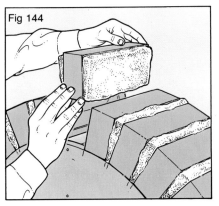

Fig 144

Fig 145

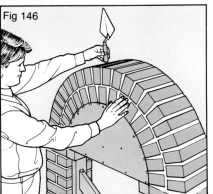

Fig 146

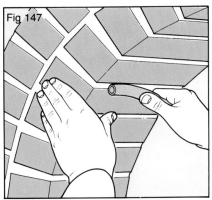

Fig 147

Fig 144 Complete the inner ring by adding the keystone.

Fig 145 Use your spirit level to check that the face of the arch is flat and truly vertical.

Fig 146 Build up the second ring in the same way after spreading a bed of mortar on top of the inner ring.

Fig 147 Neaten all the visible pointing, and leave the former in place for forty-eight hours to allow the mortar to harden. Then remove it carefully and point up the underside.

Laying Concrete Features

Concrete is one of the most versatile building materials available to the do-it-yourselfer. Not only is it an essential ingredient of many building projects, in the form of strip or raft foundations set in the ground to support walls and other structures. It is also a constructional material in its own right, and can be used to create many outdoor features such as patios, paths and drives.

The main drawback with concrete is without a doubt its utilitarian appearance. This obviously does not matter when it is used for something that is largely hidden, such as a foundation slab, but where the material is on show its looks become more important. There are two ways in which the appearance of large expanses of concrete can be significantly improved, even on a DIY basis: colour and surface texture.

Colour can be affected to a certain extent by careful choice of the sand used as part of the formula, and more drastically by the use of pigments, while the finish given to the slab can add a strong element of visual interest to the project.

What to do

In principle, laying concrete in the form of a patio, path or drive is little different from casting a slab foundation (*see* pages 26–8). However, there are several specific points to bear in mind over and above the straightforward casting technique.

First of all, if you intend to use colour in your concrete mixes, you must be prepared to carry out extremely accurate measuring of the ingredients – particularly where pigment is concerned. Since you will be using several batches of concrete for a large project, slight differences in shade will be extremely noticeable and impossible to correct.

You may want to create shapes rather more elaborate than straightforward rectangles and squares. Fortunately, concrete can do this easily so long as you are prepared to spend some time setting out the formwork in the shape you require.

As you plan the layout of your project, watch out for obstacles such as manhole covers and drainage gullies. You will need

What you need:
- pegs and stringlines
- tape-measure
- builders' square
- spade
- formwork and pegs
- hammer and nails
- spirit level
- hardcore
- concrete
- concrete mixer if not using ready-mix
- wheelbarrow
- tamping beam
- float, broom or other finishing tools

CHECK
- that you use a stronger-than-usual 1:1½:2½ formula if you are mixing your own concrete.
- that you lay a slab of the right thickness – 75mm (3in) for paths, 100mm (4in) for patios and drives.

TIP
Use ready-mix concrete for any job needing more than about 3cu m of material.

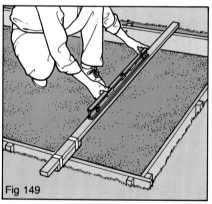

Fig 149

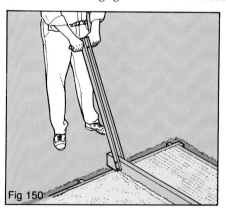

Fig 150

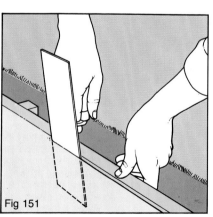

Fig 151

Fig 152

Fig 149 Set out your formwork with care, checking that you incorporate a fall across the surface.

Fig 150 Place and compact the concrete as for a slab foundation, working it well into the corners of the formwork.

Fig 151 As the concrete hardens, run the blade of a steel float round the edge of the slab.

Fig 152 For wide slabs adjoining a building, lay the concrete in alternate bays so you can use a tamping beam across the width of individual bays. Fill in the others later.

Laying Concrete Features

to plan the levels of your new surfaces carefully unless you are willing to move or reposition the obstacle.

Large areas of concrete cannot be laid as continuous slabs, or they will crack due to expansion and contraction. This means dividing the work up into the bays, each separated from its neighbour by an expansion joint of hardboard or bituminous felt if the concrete is laid as a continuous operation. If it is laid in alternate bays (see Fig 152), board or felt joints are not needed; a simple butt joint will suffice between the first set of bays and those filled in later. In either case, the recommended maximum size of each bay is around 4 × 2m (13ft × 6ft 6in). On narrow paths less than 2m (6ft 6in) wide, joints should be incorporated every 2m.

As far as finishes for concrete slabs are concerned, you have several options. You can create a smooth finish with a steel or wooden float, or use soft brooms or stiffer brushes to texture the surface. You can scatter fresh aggregate and tamp it into the surface, or score patterns in it with a wide range of implements.

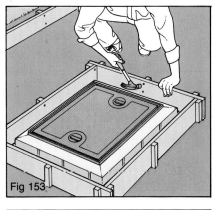

Fig 153

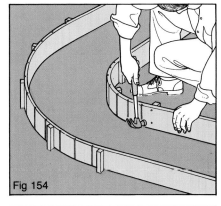

Fig 154

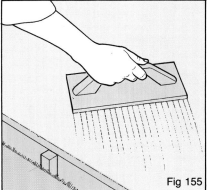

Fig 155

Fig 156

Fig 157 (*above*) Concrete is the ideal material for casting steps as well as paths.

Fig 153 Do not concrete right up to inspection chambers. Instead, cast a separate box of concrete round them, then fit expansion joints between the 'box' and the rest of the concrete.

Fig 154 Use partly sawn boards or steel formwork to create curved shapes.

Fig 155 Use a wood float to create a sandpaper texture.

Fig 156 Use a stiff broom to create a corduroy texture. If you water and brush the partly-set surface, you will leave the stones exposed and standing slightly proud of the slab surface.

Building Planters and Creating Ponds

Planters

A popular way of bordering patios and paths is to create raised planters – in effect, flower beds contained within dwarf walls of brick or stone so that the surface of the bed is some distance above ground level. Apart from their attractive appearance, raised planters are also a tremendous boon for the less physically able gardener, because they do away with the need to work on your knees or to bend a stiff back over-much.

What to do

If the planter is built on virgin soil, it should have its own foundation strip. This need be no more than 100mm (4in) wide and 100mm (4in) deep for a low planter up to about 600mm (2ft) high; for bigger structures, follow the guidelines for garden walls (see pages 42–3).

If it is being built on an existing concrete or paved surface, there is no need for extra foundations (although you should restrict the height to about 600mm (2ft) unless you know what foundations are below).

Build up the planter walls course by course, incorporating weepholes between the bricks near the base of each side if you are building on a hard surface, to prevent the planter from becoming waterlogged.

When you have completed the walls, paint the interior with a liquid damp-proofer (see below) to stop efflorescence from staining the brickwork, and then fill the base of the planter with some hardcore and gravel, topped with soil.

Fig 158 Cross-section of a planter.

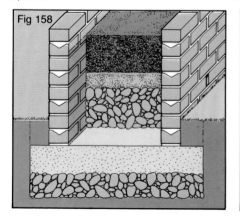

Fig 158

Ponds

A pond is the perfect finishing touch for any garden, providing the restful combination of water with the opportunity to keep fish and grow a range of attractive aquatic plants. It will also be very popular with the local wildlife.

In the old days, creating a garden pond meant casting a concrete shell and then attempting to keep it watertight as the years went by. Nowadays you have two far simpler choices: you can either buy a preformed glass fibre pond shell and set it in a matching hole in the ground, or use a flexible pond liner. The pond shells come in a fairly restricted range of designs up to a maximum length of about 3m (10ft) or so, while liners come in sizes up to 7 or 8m (around 25ft) across and can be laid in almost any shape you want. Glass fibre is virtually indestructible; liners wear out eventually, although the life of the more expensive butyl rubber liners is put at approaching fifty years. Cheap PVC liners can fail in less than five years.

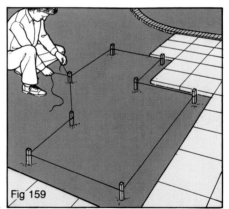

Fig 159

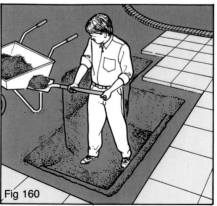

Fig 160

What you need:
- rigid pond shell or flexible pond liner
- garden spade
- spirit level and long timber straight-edge
- tape-measure
- soft sand
- garden hose
- handyman's knife
- edging stones
- mortar
- bricklaying trowel

CHECK
- that the perimeter of the hole is dead level all round. If it is not, the pond will look very odd when filled to the brim with water!

Fig 159 For a liner pond, start by marking out the pond area with pegs and string or a length of hose.

Fig 160 Excavate the hole, saving topsoil for use elsewhere in the garden, and remove sharp stones and roots.

Fig 161 Bed rigid pond shells on damp sand within a carefully-shaped hole and then lay paving slabs or bricks round the edge.

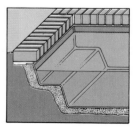

Building Planters and Creating Ponds

What to do

Start by choosing the type of pond you want to install. For pond shells, you need to excavate a hole that is a close match to the pond profile; keep testing until you get a good fit, then line the hole with damp sand and bed the shell in place.

For liner ponds, excavate the site to the size and shape you want, aiming to make the centre of the pond at least 450mm (18in) deep if you plan to keep fish so they can survive if the pond freezes over in winter. Remove all sharp stones, and cut back any plant or tree roots within the area. Then line the hole with damp sand and drape the liner into place.

Weight the edges of the liner down with stones and start to fill it using your garden hose. Carefully form creases and folds in the liner as it fills, releasing the perimeter weights as necessary. When it is full, trim off excess liner all round to leave a border about 150mm (6in) wide, and lay edging slabs or bricks round the pond on a mortar bed so they overhang the edge of the pond slightly.

Fig 166 (*above*) You can add a fountain and lighting to your pond to make it the centrepiece of your garden day and night.

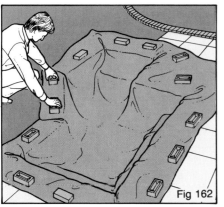

Fig 162

Fig 163

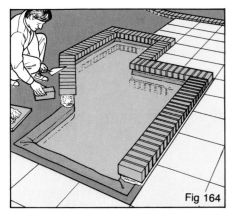

Fig 164

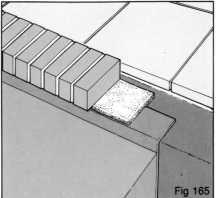

Fig 165

Fig 162 Lay the liner over the hole and weight the edges with stones.

Fig 163 Start filling the pond using a hose. Form neat folds in the liner as necessary while it fills, releasing the weights to allow it to be pulled into shape by the water.

Fig 164 Trim off excess liner and then lay perimeter slabs or bricks all round the pond on a mortar bed.

Fig 165 Make sure the perimeter paving or blocks overhang the pond edge, to help protect the liner from direct sunlight.

Fixing Fence Posts

One of the more mundane outdoor building jobs involves securing fence posts, and other vertical timbers such as pergola supports or gate posts, in the ground so they are not uprooted by the first gale of winter. If the posts fail, it is likely that the whole fence will be damaged, possibly beyond repair, so it is important to ensure that all posts are substantially anchored.

What to do

The best way of securing fence posts and similar timbers is with a collar of concrete extending from ground level to the base of the post. Dig a hole about 150mm (6in) deeper than the section of post to be embedded in the ground and a spade's width across. Ram 150mm (6in) of hardcore in the bottom and set the post in place, propping it upright while you pour the concrete in round it. Tamp it down well, and finish the top with a slope so water can run off.

An alternative method is to use a steel fence spike, driven into the ground with a sledgehammer. The post is then fitted into the socket on top of the spike.

Fig 169 (*above*) Good posts make good fencing. Make sure yours are ready for the next winter gale.

What you need:
- fence posts
- spade
- hardcore and concrete
- fence spike
- spanner
- spirit level
- scrap timber for bracing posts upright

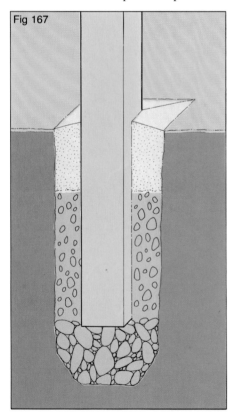

Fig 167

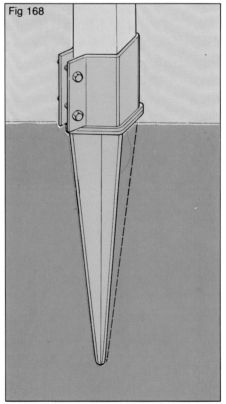

Fig 168

Fig 167 Set timber posts on hardcore, surrounded by a collar of concrete.

Fig 168 Alternatively, drive fence spikes into the ground and bolt the post into its socket.

INDOOR JOBS

Although you have more scope for building creatively out of doors, there are several jobs which you can carry out inside the house. This chapter looks at some of them.

What You Can Tackle

You are unlikely to want to do much actual building work indoors, but one job where your new-found building skills will come in useful is creating a masonry fire surround from a kit.

The most valuable indoor skill you can master is that of plastering. It is the most labour-intensive of all building crafts, and by doing your own plastering work you stand to save yourself a lot of money. Once you have mastered the technique, you will be able to tackle walls and ceilings, and to form indoor features such as arches.

A related skill is that of creating textured finishes on walls and ceilings – again a skill that is expensive to buy.

Fig 170 (*above*) Wire mesh arch formers make it easy to create arch shapes within any square-topped opening.

Fig 171 (*left*) Arches soften the look of doorways and larger openings between through rooms.

Building a Fire Surround

In the 1960s and 1970s, as central heating became more widespread there was a vogue for ripping out redundant fireplaces and bricking up the openings. Unfortunately, this often destroyed the focal point of the room, especially in main living rooms, and many home owners now want their fireplaces back so they can once again enjoy the pleasures of a real fire.

So long as the chimney breast and flue are still present, bringing a disused fireplace back into commission is not a difficult task. If the opening has simply been blocked off, all you have to do is to remove the masonry or panelling. However, if you find that the old fireback has been removed you will need to have it replaced – a job best left to a professional builder, since you need to ensure that the new fireback is correctly positioned and properly sealed to the throat of the flue so the fire will draw properly.

Once that is done you can turn your attention to reinstating the fire surround, and one of the simplest ways of doing this is to use a fireplace kit.

What to do

Once you have opened up and restored the fireplace itself, your first step is to choose a suitable kit. There is a wide range available in many different styles, consisting of a set of bricks, blocks or shaped briquettes which you assemble using mortar to construct the new surround and its superimposed hearth.

Measure the fireplace opening and the width of the chimney breast, and use manufacturers' literature to select the style you want in an appropriate size for the location. Remember that it is often possible to adapt a kit to reduce its width slightly where necessary – it all depends on the design.

While you are waiting for the kit to arrive, have the old flue swept and remove any capping that may have been fitted to the top of the flue. You may have to reinstate the chimney pot, and you should remove any air-bricks inserted into the flue for ventilation when the fireplace was originally blocked off.

What you need:
- fireplace kit
- dry ready-mixed bricklaying mortar
- wall ties, screws and wallplugs
- bricklaying trowel
- spot board
- spirit level
- stringline
- brick bolster and club hammer
- pointing trowel
- power drill plus masonry bit
- screwdriver

CHECK
- that the flue is clear and any capping or airbricks are removed
- that the existing constructional hearth is adequate to support the new surround. It should project forwards 500mm (20in) from the front of the fireplace opening, and sideways by 150mm (6in) at each side, and should be at least 125mm (5in) thick.

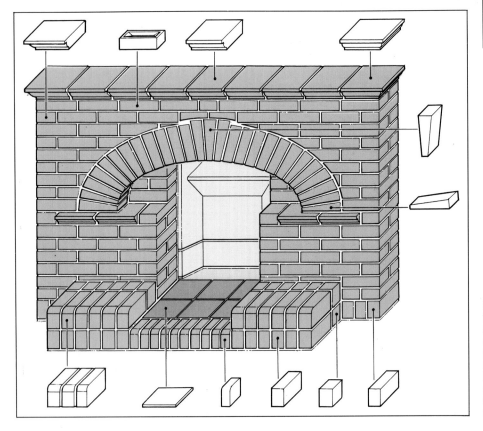

Fig 172   A typical masonry fireplace kit consists of standard and specially shaped bricks or blocks. It is wise to dry-lay the components on the floor before starting work so you can correctly identify all the components and check that none is missing.

Building a Fire Surround

When the kit arrives, check that it is complete and read the instructions carefully. If mortar is not included, it is most convenient to buy dry ready-mixed bricklaying mortar for the job. Buy mortar pigments too if you want coloured mortar; prepacked mortar can be a very drab grey colour which may spoil your surround.

Assuming that the fireback and constructional hearth are present and sound (*see* Check), you can start work. Most kits are assembled in much the same way, with the surround sides being built up first. Then you add the brickwork over the fireplace opening, and carry on up to mantelshelf level. Note that you need to tie the structure to the front of the chimney breast to ensure its stability; the kit should include special wall ties for this.

With the surround complete you can turn your attention to laying the superimposed hearth, again bedding the components in mortar. Once the surround is complete, point all the joints with a fairly dry mortar mix, taking care not to get any mortar on the faces of the bricks where it could cause unsightly stains.

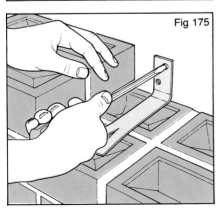

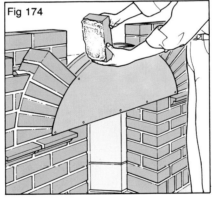

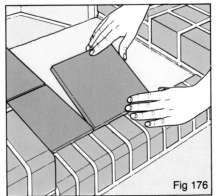

Fig 177 (*above*) The completed surround is a handsome addition to any room.

Fig 173 Start by building up the sides of the surround, checking that the courses are rising level with each other at both sides.

Fig 174 When the piers reach the level of the head of the fireplace, add the masonry across the top of the opening. The kit may include a metal lintel, or you may have to fabricate your own temporary support as shown here.

Fig 175 Secure the structure to the wall with metal ties.

Fig 176 Complete the installation by laying the superimposed hearth.

Plastering Walls

Plaster is the material most widely used for giving a smooth, hard and flat surface to internal walls, ready for decoration. Two basic types are in common use. The first is a mix based on a mineral called gypsum. The second is cement-based plaster, which is more widely used nowadays as rendering to weatherproof exterior walls, although it is still sometimes used indoors as an undercoat for other plasters. See pages 15 and 92 for more details.

Plaster is usually built up as a two-coat system, consisting of an undercoat or base coat about 10mm (⅜in) thick, covered by a finish coat about 3mm (⅛in) thick. The undercoat is designed to even out any irregularities in the wall surface, and to cope with the absorption of water by the wall; some surfaces absorb water faster than others, so there are different types of plaster for each substrate to guarantee even drying without cracking. The undercoat also evens out differences in the absorption rate between the bricks or blocks and the bonding mortar so that the thin finish coat can dry out evenly.

What to do

Assuming that you will be plastering over new brickwork or blockwork, or on a wall with the old plaster removed, the first step is to divide the wall area up into a series of bays using vertical timber battens called grounds. The purpose of these is to help you apply a flat and consistently thick undercoat to the wall, using the grounds as depth guides. They are then removed once the plaster has hardened, and the narrow channels are filled in with more plaster.

Until you become proficient at plastering, it is best to set the grounds no more than about 1m (3ft) apart. Pin them to the wall with slim masonry pins, leaving their heads protruding so you can prise them off easily later, and check that they are vertical with a plumbline or spirit level. Fit one tight into the angle of internal corners. At external corners, pin on a length of expanded metal angle bead, which will not only act as a depth guide here but will remain in place to reinforce the vulnerable corner of the wall.

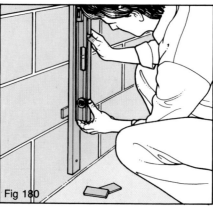

Fig 178   Divide the wall to be plastered into bays about 1m (3ft) wide, using timber battens as plastering grounds.

Fig 179   Secure each ground with partly-driven masonry nails.

Fig 180   Place packing behind the grounds if necessary to get them truly vertical.

Plastering Walls

In practice, you may find you need only two battens; once you have plastered one bay, you can remove one of the grounds and move it along the wall to form one edge of the next bay, using the plaster edge of the first bay as the ground on that side.

The first step in getting plaster onto the wall is to hold the hawk under the edge of the spot board so you can scoop some plaster onto it. Form it into a neat mound in the centre of the hawk.

Now move to the wall in front of the first bay, and wet the wall by spraying some water onto it using a garden spray gun, or by flicking water on with a large paintbrush. Load the trowel with half of the plaster on the hawk, rest the right-hand edge of the trowel on the edge of the right-hand timber ground and tilt the trowel blade up until its face is at about 30° to the wall surface (read left for right if you're left-handed). Push the trowel upwards, keeping an even pressure on the heel of the trowel where it rests on the guide batten, and letting the trowel blade move up towards the vertical as you squeeze

Fig 185 (*above*) Your new-found plastering skills will enable you to carry out a wide range of jobs.

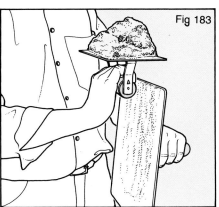

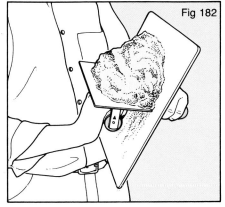

Fig 181 Wet the wall by flicking water on with a brush, or use a garden spray gun.

Fig 182 Scoop some plaster off the hawk, tilting this towards the vertical as you push the trowel across its surface.

Fig 183 Take the plaster cleanly off the now vertical hawk.

Fig 184 Press the plaster against the wall at the foot of the first bay, running the corner of the trowel up the ground as you do so.

Plastering Walls

the plaster out between it and the wall. Finish off the stroke with the trowel blade vertical as the plaster runs out, pressing its lower edge firmly against the wall. Then apply a second trowelful of plaster to the wall, immediately to the left of the first one. Work across the bay adding more until you reach the ground at the other side. Work up the wall spreading more rows of plaster dabs until you have filled the bay. You will need a simple working platform to reach the top rows.

Now use your timber straight-edge to rule off the plaster level with the face of the timber grounds. Rest its edge on the grounds at floor level and draw it upwards, keeping it roughly horizontal and sawing it from side to side so it removes any high spots. Fill in any hollows with more plaster, and rule off the surface once again. Finish off the bay by keying the surface of the plaster, using your wooden devilling float in a circular motion. Wet its base and keep it flat to the wall so the nails score shallow marks in the plaster surface.

Plaster subsequent bays in the same way to complete the first wall, then carry on round the room until you have returned to your starting point. Undercoat typically takes about two hours to harden, so you should now be able to start applying the finishing plaster.

Mix your finishing plaster in a bucket, as you did for the undercoat, and transfer it to the spot board. It should be much runnier – roughly the consistency of melting ice cream. Scoop some onto your hawk, move to the wall and apply it over the undercoat in a thin layer. Work this time from bottom left to top right over an area about 2m (6ft 6in) wide, using broad sweeping arm movements. When you have covered this area, go back to where you started and apply a second, even thinner, coat. Trowel off any ridges or splashes with light downward strokes of the trowel edge. Then wet the blade and work over the whole area again with the blade at about 30° to the surface.

Once the finish coat has hardened, wet your trowel again and polish the surface to give a perfectly smooth flat finish.

Fig 190   Pin angle beads or a timber ground to external corners.

Fig 191   Plaster the bay and then rule off the surface as before.

Fig 186

Fig 187

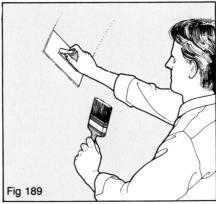

Fig 189

Fig 186   Apply more plaster in rows, working up and across the bay.

Fig 187   Then use the rule to remove high spots, and fill any hollows before ruling off again. Key the surface ready for the finish coat.

Fig 188   After the undercoat has hardened, apply the thin finish coat with sweeping arm movements across the wall surface.

Fig 189   Polish the finish coat with the trowel, wetting the surface as you work.

Plastering Ceilings

If you have replaced an old ceiling with new plasterboard, you can give it a finish coat of plaster ready for decorating. However, because the boards can move as temperature and humidity changes affect the joists above the ceiling, it is essential to reinforce all the joints between the boards and round the perimeter of the room so they do not crack, using scrim tape bedded in plaster.

Fig 192

What to do

Start by spreading a thin band of Thistle board finish along all the joint lines. Then bed one end of a 75mm (3in) wide band of fine scrim into the wet plaster, and use your trowel to press it into place along the seam. Use a similar technique to bed more scrim into the wall/ceiling angle all round the room.

Next, spread a thin coat of the finish plaster along all the scrimmed joints, and start to plaster the 'bays' between the joints, putting on a thin coat of plaster to match the thickness of the scrimmed joints. When this is complete, go back to your starting point and apply a second thin coat over the entire ceiling surface.

Rule off the surface with an aluminium straight-edge (called a darby) to remove any irregularities, and neaten the angle between ceiling and wall by drawing your trowel along the ceiling with one corner trimming the plaster evenly. You can use an angle trowel here if you prefer. Finish the plastering by polishing the surface with the wetted blade of your trowel.

What you need:
- Thistle board finish
- scrim tape
- bucket and stick
- spot board
- plasterer's trowel
- hawk
- aluminium straight-edge
- angle trowel
- scissors for scrim
- water brush or garden spray gun

CHECK
- that you have a safe working platform so you can reach the ceiling comfortably. Boards set on trestles (see page 30) or a low mobile trolley made from platform tower components are ideal.

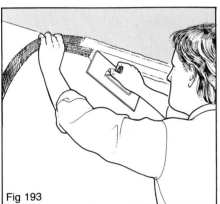

Fig 193

Fig 194

Fig 195

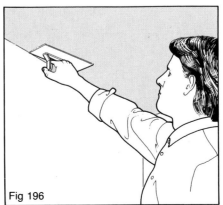

Fig 196

Fig 192 Spread Thistle board finish along all the joint lines, then bed scrim into it with your trowel.

Fig 193 Apply scrim to the wall/ceiling angle too to prevent cracks along the joint line.

Fig 194 Spread more finish plaster along all the joint lines to hide the scrim.

Fig 195 Apply the first thin coat of plaster to the exposed board surface between the joint lines. Follow it with a second coat, ruled off neatly with an aluminium straight-edge.

Fig 196 Neaten the wall/ceiling junction by running the corner of your trowel carefully along the angle.

Creating Arches

If you want to relieve the plainness of a room by introducing arch shapes within doorways or across converted through-rooms, the simplest way of doing so is to use prefabricated arch formers. These are made from expanded metal mesh, assembled into arch shapes with flat faces and curved undersides. Each end of the arch consists of two sections, (a vertical face panel attached to a curved soffit panel), and you offer one up against each side of the opening so the soffit sections overlap and form a continuous curved underside. Because of this, one set of formers can be used to form arches in walls of different thicknesses.

The simplest kits consist of formers to create quarter circles at each side of the opening (or a semi-circle in doorways). More elaborate versions allow you to create full-width flat or oval arches, or exotic types such as Tudor, Spanish or Arabian arches. Some full-width arch kits also have extra sections to fill out the centre of the arch – a total of six in all, three for each face of the arch.

What to do

Start by selecting the arch shape you want, and then buy a kit that will cater for the overall width of the opening you intend to bridge. Check also that it can cope with the thickness of the wall if this is more than about 200mm (8in).

Offer up each of the end sections in turn to the wall so you can mark its outline on the surface. Use a club hammer and sharp brick bolster to cut back the plaster as necessary. Then fix the former sections in place with masonry nails or screws and wallplugs, and add any central sections on wide arches. Wire the overlapping soffit sections together with galvanized wire. Tuck the ends into the mesh so they will not protrude through the plaster. Check that all the sections are correctly aligned.

Then start applying plaster to the soffit section, using the curved angle beads as a thickness guide. When that is covered, rule it off if necessary and plaster the side sections. When the undercoat has dried, float on a thin finish coat.

What you need:
- arch former kit
- metal lathing plaster
- finish plaster
- brick bolster and club hammer
- masonry nails or screws and wallplugs
- fine galvanized wire
- claw hammer
- wire cutters
- bucket and stick
- spot board and hawk
- plasterer's trowel

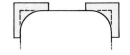

Fig 201   Arch shapes.

Fig 197

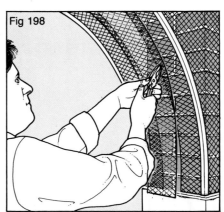

Fig 198

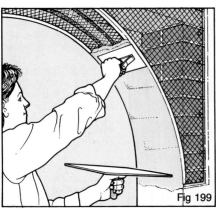

Fig 199

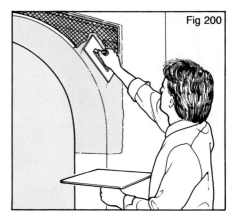

Fig 200

Fig 197   Cut back the plaster as necessary and nail the former sections in place.

Fig 198   Tie the curved soffit sections together with galvanized wire.

Fig 199   Plaster the underside of the soffit.

Fig 200   Then plaster the sides and add a finish coat overall.

Creating Textured Finishes

Textured finishes are an attractive alternative to relief wallcoverings if you want a three-dimensional surface to your walls and ceilings. They are also an excellent cover-up for less-than-perfect plasterwork on which you do not want to spend a lot of preparation time before painting the surface in the usual way.

There are two main types of textured finish available. The first comes in powder form; you simply add water to make up a creamy mix which you apply to the surface you are decorating, ready for texturing. Artex is the most widely available brand; indeed, the name has become synonymous with textured finishes in the same way as Formica means plastic laminate for kitchen worktops. It is now also available in ready-mixed form.

The second finish is better known as textured emulsion paint, and gives a less pronounced relief finish to the surface. It is ready to apply straight from the tub.

Most of these finishes dry white or off-white, and can be painted using emulsion if you want a coloured finish.

What to do

You can apply textured finishes to any wall or ceiling surface so long as it is properly prepared. That means washing down existing painted surfaces with sugar soap or strong detergent, removing any flaking material and filling cracks larger than a hairline with general-purpose filler. Fill and tape all joints in new plasterboard. You must remove wallcoverings completely. Then prime new plaster or plasterboard, or bare plaster exposed after wallcoverings have been stripped, using a proprietary plaster sealer. This will stop the textured finish drying too quickly and cracking as it does so because of excess porosity.

With powder types, mix the contents of a 5kg (11lb) bag with 2¼ litres (4 pints) of water, leave to stand for ten minutes and then add a further ½ litre (1 pint) more water. Stir thoroughly. Ready-mixed products can be applied direct from the tub.

Apply powder and thick ready-mixed types with a wide paintbrush; use a roller if you prefer with textured emulsion paints.

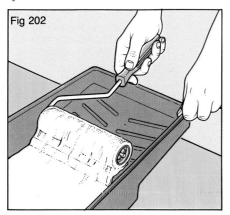

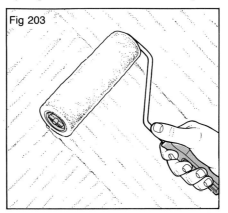

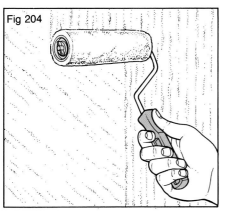

Fig 202 You can apply textured emulsion using a paint roller. Half-fill the tray and load up the roller.

Fig 203 Apply the coating fairly thickly. Use a brush with powder types and the thicker ready-mixed ones.

Fig 204 Even out your brush or roller marks by working over the coated surface in one direction only.

Fig 205 Use a narrow brush to form a neat band round obstacles and at ceiling edges.

Creating Textured Finishes

Cover an area of about 2sq m (22sq ft) at a time, then texture the freshly-applied coating with whatever tools you have decided to use. At wall edges and round obstacles, use a narrower brush to form an untextured band that will 'frame' the textured area neatly.

Creating Patterns

You can use a wide range of proprietary and home-made implements to create your patterns. A plain short-pile paint roller will form a uniform fine stipple effect, while a special texturing brush can be used to form a coarser stipple, or can create circular and figure-of-eight swirls. You can buy roller sleeves with a range of embossed designs and cut-out sections to form bark patterns or regular relief designs. Toothed combs can also be used to create your own regular repeating patterns. You can even slap the finish on with a flexible spatula to create the well-known Spanish restaurant effect of randomly-applied plastering. The secret of success is to practise the effect you want *before* starting work in earnest.

Fig 206 (*above*) The effects you can achieve with textured coatings can be highly dramatic.

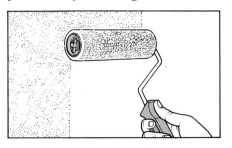

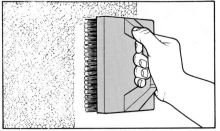

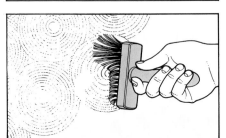

Fig 207 You can use a wide range of brushes, roller sleeves and other tools to create textures and patterns on wall and ceiling surfaces. Experiment first on scrap board to gauge the look of a particular effect and to master your technique.

MAINTENANCE & REPAIRS

We all expect our homes to be weatherproof, and most of the time the average house does its job pretty well. But a house is a complex structure, with all sorts of inherent weak spots, and it only needs one small component to fail before rain gets in, draughts spring up or frost gets to work. Left unattended, these small problems can grow into big ones, and may cost a lot of money to put right.

The answer is to give your house a regular check-up – rather like a car service – so you can put any faults right before they can cause serious trouble. So get out the clipboard and pen, and take a quick guided tour. Below are some of the key parts of the structure to put on your outdoor checklist.

There are also several repair jobs you are likely to need to carry out indoors, from patching damaged plasterwork to levelling concrete floors.

An Outdoor Checklist

If your house has any chimney stacks, there are three main problems to look for. All can allow water penetration, causing damp patches to appear on ceilings and chimney-breasts. To carry out your inspection, either climb a securely-roped extension and roof ladder, or climb just to eaves-level and use binoculars.

The first thing to check is the state of the sloping mortar cap (called flaunching) that holds the pot in place. If this is cracked or lifting, it needs hacking off and replacing with fresh mortar – a job for a local builder if you are not happy working at such heights.

Next, check the state of the pointing on the stack. If it is crumbling or missing in places, repointing is necessary. It may also be worth planning to give the stack a couple of coats of clear silicone masonry sealer to make it waterproof and to help resist frost damage.

Lastly, check that the metal flashings which waterproof the join between the stack and the roof slope are secure and well dressed down onto the tiles. Strong winds often lift and tear them, failed pointing allows them to pull away from the stack, and age can make them porous.

As for roof repairs, look for loose tiles, slipped slates and signs of damage to flat roof surfaces. If you are happy working on the roof you can tackle these yourself; otherwise call in a builder to do the work for you.

While you have your ladders out, take the opportunity to check the gutters. Lift out any debris that has collected, using a garden trowel or an empty food tin as a scoop. Then wash them through with water from your garden hose, so you can see if there are any parts that need attention; leaks and overflows can lead to penetrating damp through the walls below.

Apart from the roof, it is the house walls that keep most of the weather at bay . . . and suffer as a result. If you have exposed brickwork, look out for crumbling or missing pointing and signs of frost damage, which causes the face of the brick to split (called spalling) and can allow damp to penetrate.

If you have rendered or pebble-dashed walls, check for loose patches by tapping the wall surface with a screwdriver handle or similar implement. Any you find should be cut out and re-rendered; otherwise they will eventually just fall away.

Finally, check the condition of ground-level areas such as paths, drives and steps, looking out for cracks, potholes and uneven slabs that could be dangerous.

Fig 208 (*above*) Every home needs regular maintenance and repair work to keep it weatherproof and in good condition.

Repointing Brickwork

If your house has external walls of fair-faced brickwork, it is at risk from a gradual deterioration of the mortar joints between the bricks. As time goes by, rain can erode the mortar and exaggerate any weaknesses in the original work – the wrong mortar mix, for example, or simply poor workmanship when the wall was built. Once water starts to penetrate at the joints it is only a matter of time before it starts to soak into the facing brickwork. Then frost can cause the tell-tale splitting of the brick face known as spalling, and you are heading for some expensive repair work.

It is therefore worth inspecting the condition of your exterior brickwork at regular intervals, so you can pinpoint and repair any areas that are showing signs of decay. Look especially for areas where the top edge of the bricks have become exposed due to failure of the joint, and use a tool such as an old screwdriver to dig into pointing that appears to be crumbling or loose at any point. Localized repairs now may avoid the need for extensive wholesale repointing later.

What to do

When you have identified areas that need some repair work, your first step is to hack out the failed mortar to a depth of about 19mm (¾in), using a narrow cold chisel and a club hammer. Work along the horizontal bedding joints first, then tackle the vertical joints between them. Brush out loose material from the joints.

The secret of success with repointing is to match not only the type of joint used on the wall, but also the colour of the mortar used. Matching the joint is just a matter of careful workmanship, but matching the colour can be more difficult, especially if the masonry is heavily stained. Experiment with pigments, allowing trial mixes to dry so you can compare the colour with that of the existing pointing. It may be worth cleaning badly discoloured brickwork first (see page 73).

When you are ready to start work, wet the joints and bed the mortar firmly into place with a pointing trowel. Do the horizontals first, then the verticals, and finish off by matching the joint profile.

What you need:
- sharp cold chisel
- club hammer
- stiff brush
- water brush or garden spray gun
- bricklaying mortar
- pointing trowel
- hawk and spot board
- profiling tools

CHECK
- that you use the right mortar formula if you are mixing your own. Use 1:1:5 cement:lime:sand, 1:4 masonry cement:sand or 1:5 cement:sand with added plasticiser.

TIP
Do not make up more mortar than you can use in an hour (remember that repointing is slow work). Mix small batches so you can measure out all the ingredients accurately – essential if you are using pigments.

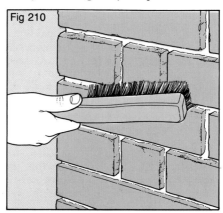

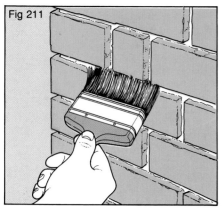

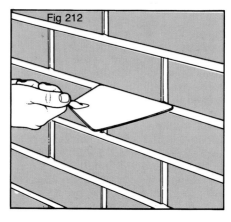

Fig 209   Use a cold chisel and club hammer to hack out the old mortar to a depth of about 19mm (¾in).

Fig 210   Brush all loose material out of the joints with a stiff brush.

Fig 211   Wet the joints just before you start repointing them to cut the suction rate.

Fig 212   Repoint the horizontal joints first, then the verticals, and finish the joint profile to match the rest of the wall surface.

Patching Rendering

Many homes have exterior walls which are finished with rendering of one sort or another. This is often applied to provide additional weatherproofing either when the wall was first built, or at a later stage. It is a layer of cement mortar, spread and levelled on the wall surface in much the same way as walls inside are plastered. The finished surface may be left smooth, or may be textured slightly; it may even have pebbles or other finely-crushed aggregate embedded in it to form a high relief surface finish.

As time goes by, slight movement in the masonry coupled with the effects of sun, wind and rain eventually cause the rendering to crack. Water can then penetrate and seep down between the rendering and the masonry, and frost can cause the bond between the two to break. The result is areas of rendering which are hollow when tapped, and which can eventually break right away from the wall to expose the bare masonry behind. These patches need prompt repair before larger areas are affected.

What to do

To locate areas of failed rendering, look out for fine cracks in the surface and then tap the rendering around them with the handle of a screwdriver or some similar tool. Mark any areas that sound hollow with chalk.

Next, use a club hammer and a sharp brick bolster to cut away the loose areas until you reach a sound edge. Undercut the edges of the area slightly so the patch has a good key with the old material, and brush out all loose debris.

Mix up some repair mortar and carry it to the wall. Use dry ready-mixed rendering mortar for small-scale repairs; for larger jobs use the same mix as for repointing (see page 70) unless the wall is moderately sheltered, when a weaker mix can be used (see Check).

Wet the wall surface first, then use a plasterer's trowel to fill the patch with mortar. Rule off any excess material with a timber batten, fill any hollows and then texture or pebbledash the surface as required to match the surrounding area.

What you need:
- sharp brick bolster
- club hammer
- stiff brush
- water brush or garden spray gun
- rendering mortar
- plasterer's trowel
- hawk and spot board
- texturing tools

CHECK
- that you do not use too strong a mix in sheltered conditions. Here you can use 1:2:8 cement:lime:sand, 1:5½ masonry cement: sand or 1:7 cement: sand with added plasticiser. The sand should be clean washed plastering sand.

TIP
For added strength, brush diluted PVA building adhesive onto the masonry before filling the damaged area.

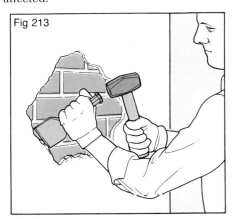

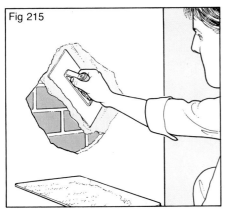

Fig 213 Once you have identified weak areas, hack off the old material with a brick bolster and club hammer.

Fig 214 Wet the masonry surface to reduce suction.

Fig 215 Trowel fresh mortar onto the area you are patching.

Fig 216 Use a timber batten to rule it off level with the surrounding rendering.

Patching Concrete Sills

Precast concrete sills were very popular in Victorian and Edwardian homes as a cheap alternative to solid stone, and may also be found on more recent properties. Door thresholds can suffer from physical damage as time goes by, while both door and window sills are also prone to weathering which can allow water to penetrate the concrete. Frost may then cause cracks and splitting of sections of the concrete, leaving the surface crumbling, unsightly and in the case of door thresholds, dangerous too.

It is possible to buy replacement sills, but the work involved in trying to remove the old sill and bed the new one in its place is seldom justified. It is usually simpler to attempt to repair the existing sill *in situ*, using mortar or fine concrete as the repair material. This can then be disguised by giving the newly-repaired sill a couple of coats of masonry paint.

You can use a variation on the repair technique shown here to patch similar damage where it has affected the edges of the treads on concrete steps.

What to do

Start by assessing the extent of the damage by brushing and scraping away all loose and flaking material. Then check along all cracks to see whether larger chunks of the concrete are loose, and prise them away with an old screwdriver or similar tool if they are. Use a cold chisel and club hammer to remove stubborn but cracked sections, and undercut the edge of the damaged area so the new concrete has a better key.

For small-scale repairs, use either a strong mortar mix (*see* Check) or an exterior-quality filler to build up the patch so it matches the level of the surrounding masonry.

Where larger sections of the sill are missing, it is better to make up a timber mould to contain and support the concrete while it hardens. With window sills, use timber brackets or props to hold the mould in place. Then fill in the repair and leave it to harden for up to seventy-two hours before removing the formwork carefully.

What you need:
- cold chisel
- club hammer
- stiff brush
- exterior filler *or* bricklaying mortar *or* fine concrete
- small bricklaying or pointing trowel
- hawk and spot board
- timber for mould

CHECK
- that you use a strong mix for mortar repairs: 1:½:4: cement:lime: sand or 1:2½ masonry cement:sand or 1:3 cement:sand with added plasticiser.

TIP
Brush diluted PVA building adhesive onto the damaged area before filling it, to improve the adhesion of the repair material.

Fig 217

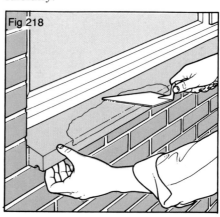

Fig 218

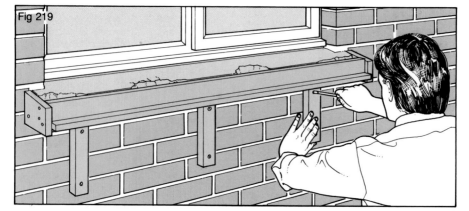

Fig 219

Fig 217   Use a cold chisel and club hammer to hack away all loose or crumbling material, then brush off any dust and debris.

Fig 218   For small-scale repairs, shape filler or mortar to match the sill profile.

Fig 219   For larger-scale jobs, make up a timber mould to support the repair material while it hardens. Leave it in place for seventy-two hours so the repair can harden fully.

Cleaning Masonry

Masonry surfaces become stained and dis-coloured for a variety of reasons, although two of the commonest are atmospheric pollution and algal growth. Especially in industrial areas, exterior walls become coated with what seems like deeply-ingrained dirt, while horizontal surfaces such as paths and drives simply get grubby from regular use. Algal growth flourishes on surfaces that are perpetually damp and receive little or no sunshine – north-facing walls, for example. Lastly, surfaces may suffer from localized stains – mortar marks, for example, or paint splashes. All can be cleaned up by one means or another.

What to do

How you tackle the problem depends on what you are up against. One of the most versatile pieces of cleaning equipment around is the high-pressure water spray gun, which can shift dirt and algal growth from walls and ground-level surfaces of all types. You can hire the equipment for as long as you need it from plant hire firms; all it needs is a water and power supply.

For more drastic cleaning, you can use a heavier-duty version of the water gun to blast abrasive particles at the wall, but this is generally best left to a professional operator.

You can tackle mortar stains with an acidic masonry cleaner, which dissolves the stain and leaves the surface clean and bright. Use them with care.

Lastly, you can generally remove paint splashes from masonry surfaces using repeated applications of the appropriate chemical paint stripper.

Fig 222 (*above*) Green algal growth and flaking paint must be scrubbed off before redecorating.

What you need:
- high-pressure water spray gun (hired)
- masonry cleaner
- paint stripper

TIP
Always angle the jet from a high-pressure spray gun away from you, and wear safety goggles when using one to guard against water and debris splashing into your eyes.

Fig 220

Fig 221

Fig 220 Use a high-pressure spray gun for washing down walls prior to repainting . . .

Fig 221 . . . or for cleaning up all sorts of paved surfaces around the house.

Patching Paths and Drives

The various hard surfaces around your property – patios, paths, driveways and the like – are always vulnerable to weathering and accidental damage. Frost is one of the biggest enemies, freezing water that gets into cracks in the material used for constructing them and eventually breaking sections away from corners and edges. Localized cracking can be caused by physical damage – overloading, for example – or by underground subsidence, and as material crumbles away from the edges of the crack a pothole can develop. This in turn lets more water penetrate, causing more widespread damage that can eventually lead to the complete break-up of the surface concerned.

For hard surfaces laid with 'seamless' materials such as concrete or tar macadam, it is relatively easy to make localized repairs which will prevent further deterioration of the surface, although it is often difficult to disguise the fact that a repair has been made – the repair shows up as exactly that unless you are fortunate in matching your choice of repair material to its surroundings.

What to do

With **concrete** surfaces, start by surveying the extent of the damage. Rake out cracks with an old screwdriver or similar tool, and cut back the edges if they are breaking away, using a cold chisel and club hammer. Enlarge the area until you reach sound material, and leave an undercut edge so the repair patch can bond well to the surrounding concrete. Where a pothole has formed, dig or pick out all loose material and then pack in some hardcore to form a solid base for the repair.

At the edges of concrete slabs, cut back loose material and then place a timber batten against the edge to act as formwork and support the repair material.

Fill wide cracks, potholes and broken-down edges with a strong concrete mix (see Check). Pack the material in well and level it with the surrounding concrete, texturing the surface as appropriate to get as good a match as possible.

To repair **tar macadam**, simply dig out loose material and repair the damage with well-compacted cold-roll asphalt.

What you need:
- cold chisel or brick bolster
- club hammer
- stiff brush
- concrete or cold-roll asphalt
- timber for formwork
- steel float or pointing trowel
- hawk and spot board
- punner for asphalt

CHECK
- that you use a strong concrete mix for repair work to in-ground slabs – 1:3½ cement: combined aggregates is ideal.

TIP
Brush diluted PVA building adhesive along cracks and round the edges of potholes to improve the adhesion of the repair concrete.

Fig 223

Fig 224

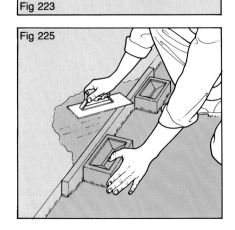

Fig 225

Fig 226

Fig 223 Dig out loose material as necessary and pack in hardcore to form a solid base for the repair.

Fig 224 Fill the patch with concrete, smooth it down level with its surroundings and then texture the surface to match.

Fig 225 At the edges, use a timber batten to act as formwork and support the repair material while it hardens.

Fig 226 Fill holes in tar macadam by ramming in some cold-roll asphalt with a punner.

Levelling Uneven Paving

Unlike concrete and tar macadam, paved surfaces have a host of seams; every joint between adjacent paving units is a potential weak spot, whether it is pointed with mortar or just butt-jointed and filled with sand. Pointed joints suffer from frost damage, while both types can be opened up by ground movement, by subsidence of the base on which they were laid, by overloading due to heavy traffic on the surface, even by weeds growing through them. The end result is not only unsightly in terms of the paving's appearance; it can also be dangerous as raised edges act as trip hazards.

The cause of the problem can often be traced back to inadequate preparation and improper techniques during the original laying of the paving. For example, crazy-paving laid on sand will always subside eventually, even if the joints are fully pointed, while block paving can settle by different amounts at different points if it has to tolerate localized overloading. The solution to the problem is to lift affected areas and re-lay the individual paving units.

What to do

Start by surveying the extent of the problem. With dry-laid units – slabs or paving blocks on a sand bed – the likeliest problem is subsidence, although with slabs individual stones may also be cracked. With pointed units, failure of the pointing is the commonest problem so long as the units were originally laid on a continuous mortar bed over a suitably stable sub-base. Use a long straight-edge to check for subsidence over larger areas.

When you have identified the problem, you can tackle the cure. Lift loose-laid units and add sand or mortar dabs beneath them to bring them back level with their neighbours. If you cannot raise an individual unit in the centre of a paved area, you may have to lift others first, working in from the edge of the area until you reach and can lift the worst affected unit.

Where pointing has failed but the units are otherwise stable, rake out the old mortar and repoint all the joints with fresh mortar. Rebed any individual stones that have subsided or are loose.

What you need:
- long straight-edge
- spirit level
- brick bolster for lifting units
- sand or bedding mortar
- bricklaying trowel
- spot board
- pointing mortar
- pointing trowel

CHECK
- that re-laid paving still has a slight fall in one direction across the area to prevent water from collecting on its surface.

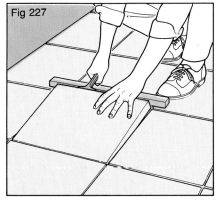

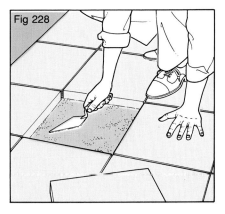

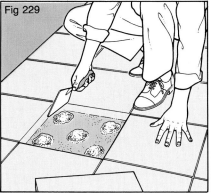

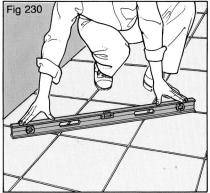

Fig 227   Identify slabs that have subsided, and lever them up with a brick bolster over a timber fulcrum.

Fig 228   Add extra sand to raise the slab level with its neighbours . . .

Fig 229   . . . or place dollops of mortar on the sand bed.

Fig 230   Replace the slab and tamp it down so it is level with the surrounding paving.

Fixing Loose Chimney Pots

Chimney stacks are a potential weak spot in the weatherproof exterior of the house, for two reasons. Firstly, the stack itself can suffer from defective brickwork, failed pointing or an insecure pot, all of which can allow water to penetrate the structure and give rise to damp patches on chimney-breasts and ceilings inside the house. Furthermore, because the stack breaks through the roof slope, the junction between the roof and the stack must be kept waterproof, and the complex flashings needed to do this can deteriorate with age or may be damaged by high winds.

The first step in assessing what needs to be done is to make a visual inspection of the stack. You can do this from ground level with binoculars, but it is better to examine it at least from eaves-level by putting up your ladder with a ladder stay as close to the stack as possible. If you are happy at heights and the design of the roof allows it, you can put up a roof ladder alongside the stack and carry out your inspection at close quarters. If you are not, leave it to an expert to do the work for you.

What to do

If you are carrying out the work yourself, it is safest to set up a roof ladder and then to use the components of a platform tower to build a working platform round the stack. You will then have a base on which to store tools and materials, plus the added safeguard of a handrail and toe boards all round. Never emulate professional roofers and try to climb directly on the roof slope; one slip could kill.

Start the repair work at the top. If the pot is loose and the sloping mortar (called flaunching) securing it to the stack is cracked or missing, secure the pot to the stack with rope. Then chop away the old flaunching with a brick bolster and rebed the pot in fresh mortar, using a 1:5 mix with added plasticiser. Use pieces of slate or fibre cement sheet to support the pot if necessary, and slope the flaunching down from the pot to the edge of the stack to assist rainwater run-off.

If the stack pointing is defective, rake out the joints and repoint with 1:5 mortar, finishing off the joints neatly.

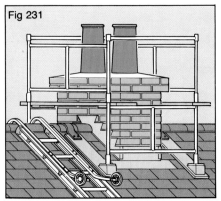

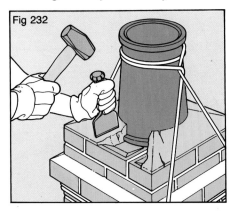

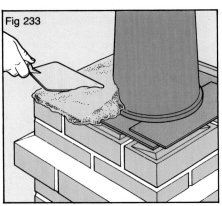

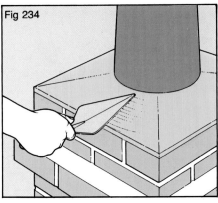

Fig 231   Before carrying out any work on the stack, it is safest to erect a solid working platform around it using slot-together platform tower components. Fit toe boards and a guard rail all round.

Fig 232   Rope the pot to the stack so it cannot move, and chop away defective flaunching with a brick bolster and club hammer.

Fig 233   Support the pot on slates if necessary, then remake the flaunching with 1:5 repair mortar.

Fig 234   Finish the flaunching smoothly with a slope to encourage rainwater to run off.

Repairing Faulty Flashings

Where flat or lean-to roofs meet a house wall, the junction between roof and wall is weatherproofed by a flashing strip. The best flashing material available is lead sheet, and this is still used on all top-quality building work. Its main advantage is that it can be easily shaped to conform to the profile of the roof surface, and it is extremely durable, although it can be lifted and torn by high winds. Felt flashings are less durable, being prone to cracking and splitting, but are often used as an economical alternative to lead on flat roofs. Mortar flashings are the least satisfactory of all materials because they soon crack due to differential movement between wall and roof, and can quickly become porous or break away completely.

When flashings fail, water can penetrate at the junction of roof and wall and may then run down roof timbers to emerge at a point indoors remote from the source of the leak. For this reason, it is worth inspecting the adjacent roof surfaces for faults at the same time that the flashing repairs are carried out.

What to do

Where **lead flashing** has been torn from its chase but is otherwise undamaged, it can be replaced relatively easily. Start by lifting the whole length of flashing away. Then rake out the old mortar from the length of the chase, saving any wedges of lead used to hold it in place, and offer the strip back into position. Tuck the lead wedges into the chase to hold the top of the strip in place; if you have none, use slips of slate or similar material instead. Then mix up a quantity of 1:5 cement:sand mortar (or buy a bag of dry ready-mixed repointing mortar), and fill the chase, pointing up the finish neatly. Then dress the replaced flashing neatly down over the roof surface.

Felt and lead flashings that are torn or damaged should be pulled away and replaced completely with a new flashing using self-adhesive flashing tape.

Hack away all old **mortar flashings** completely, and replace them with flashing tape too.

What you need:
- cold chisel
- club hammer
- pointing trowel
- 1:5 repair mortar
- self-adhesive flashing tape
- flashing primer
- sharp handyman's knife
- wallpaper seam roller

CHECK
- that lead flashings are well dressed down onto roof surfaces to prevent high winds from lifting them.
- that self-adhesive flashing tape is firmly bonded to the wall surface, with no air bubbles trapped beneath it.

TIP
In cold weather, store flashing tape indoors for an hour before using it, to make it more flexible to handle and easier to apply.

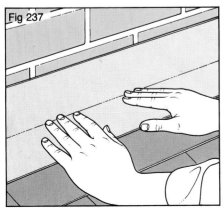

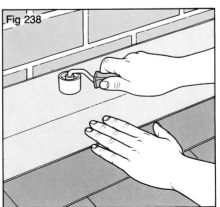

Fig 235   Remove old felt or mortar flashings and brush flashing primer onto the wall and roof surfaces. Leave it to become tacky.

Fig 236   Peel off the backing paper and press the tape into place.

Fig 237   Apply even pressure to the tape to bond it firmly to the wall and roof surfaces.

Fig 238   Use a wallpaper seam roller to ensure that you have created a waterproof seal between tape and wall.

Curing Rising Damp

Most homes built since the late nineteenth century have a physical damp barrier known as the damp-proof course (DPC) built into the exterior walls to stop ground moisture from being drawn upwards into the masonry. The DPC can fail locally, especially where materials such as slate or engineering bricks were used, allowing tell-tale semi-circular damp patches to appear low down on the inside of the wall. If this occurs, a new DPC can be injected into the masonry (*see* page 79) to cure the problem.

However if you spot rising damp do not immediately suspect DPC failure; the cause may be something bridging the DPC and allowing moisture to bypass it.

Some of the commonest DPC bridges are shown below. The first is caused by a wall or building being erected next to the house without either a horizontal or vertical DPC being included. The second results from soil or other material being piled against the wall above DPC level. The third is due to a step or other slab being built to a level higher than the house DPC without a vertical DPC between wall and slab, again allowing moisture to bypass the house DPC. The fourth is the result of rendering being applied to the wall over the DPC line.

Lastly, damp can also occur if hard surfaces are built close to DPC level and rainwater splashes up the wall above it.

What you need:
To cure damp bridges:
• strip DPC material
• silicone sealer
• paintbrush
• sundry building tools

CHECK
• that path and patio levels adjoining the house are kept at least 150mm (6in) below DPC level, so rainwater falling on them cannot splash up onto the wall above the DPC.

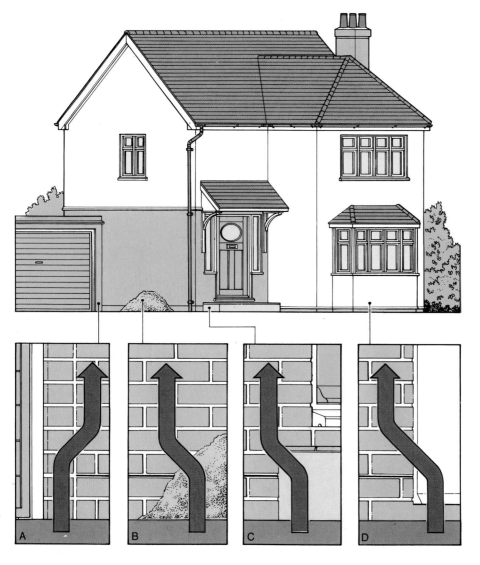

Fig 239 Four common DPC bridges. In **A** either insert a new horizontal DPC in the wall adjoining the house, or fit a vertical DPC between the two walls. In **B**, remove the soil heap. In **C**, fit a vertical DPC between slab and wall. In **D**, cut back the rendering to just above the level of the DPC.

Curing Rising Damp

What to do

Start by removing banked-up soil or other material against the house wall which could be bridging the DPC, and check that all air-bricks are clear. Where walls or slabs have been constructed so as to bridge the DPC, cut out the mortar or concrete next to the house wall and insert a vertical DPC. If paving slabs next to the house are close to the level of the DPC, either treat the house wall immediately above it with a silicone water repellent sealer or remove the slabs next to the wall and replace them with a layer of gravel to prevent rainwater splashing up the wall. Cut away rendering that bridges the DPC.

If these steps, followed by a suitable drying-out period, do not cure the damp problem, it is highly likely that the DPC has indeed failed. You can call in a specialist firm to install a new DPC, but this is a labour-intensive job and since you can hire the same pressure injection equipment that a specialist uses, there is no reason why you should not do so and inject a new chemical DPC yourself. You buy the special chemicals when you hire the equipment, on a sale-or-return basis.

Exactly how you proceed depends on whether you are dealing with solid or cavity walls. The former are usually about 230mm (9in) thick, while the latter generally measure around 280mm (11in). The injection is carried out in two stages, working wherever possible from outside the house, so the depth of the holes you drill in the masonry depends on the wall thickness. For both types of wall the first hole is drilled about 75mm (3in) deep, and the first stage of the injection is carried out. Then the same holes are drilled deeper – to 150mm (6in) in solid walls, and 200mm (8in) in cavity walls – and the second stage is completed using longer nozzles.

Once the new DPC has been injected, you can fill the injection holes with mortar coloured with pigment to match the colour of the bricks.

If the interior plaster is badly affected by damp, you will have to hack it off and replace it with a waterproofed cement render followed by a coat of finishing plaster.

What you need:
To inject A DPC:
- DPC injection machine (hired)
- DPC fluid
- power drill plus long masonry drills
- depth stop
- mortar and hawk
- pointing trowel
- brick bolster
- club hammer
- render and plaster
- plastering tools

CHECK
- that you have enough DPC fluid – you may need up to 3 litres per metre of wall if the bricks are very porous.

TIP
Hire a professional-quality power drill to make the injection holes. This avoids the risk of overloading and burning out your drill.

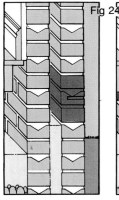
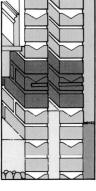
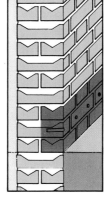
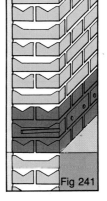

Fig 240

Fig 241

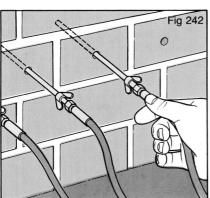

Fig 242

Fig 243

Fig 240 Inject the outer leaf of a cavity wall first, then drill through the first set of holes to a final depth of 200mm (8in) and treat the inner leaf.

Fig 241 Treat solid walls in two stages too, this time drilling the hole for the second stage to a depth of 150mm (6in).

Fig 242 Change over to longer nozzles for the second stage of the injection.

Fig 243 If the interior plaster has been badly affected by rising damp, you will have to hack it off and replace it with a coat of cement render containing waterproofing additives, followed by a finish coat of plaster.

Levelling a Concrete Floor

Many homes have some or all of their ground floors laid as concrete slabs, and these can suffer from a number of faults which may need attention before you can lay new floorcoverings or construct built-in furniture standing on them.

One of the commonest faults is dusting of the concrete surface, often caused by over-trowelling of the surface as the slab was laid. Here all you need to do is to apply a proprietary concrete sealer to the surface, or else apply a diluted coat of PVA building adhesive to bond the surface particles together.

Cracks and small potholes can appear as time goes by, the former as a result of slight ground movement beneath the floor and the latter in heavy traffic areas or where castored furniture is regularly moved backwards and forwards. Here you should rake out loose material, undercut the edges of the area to be patched and then use a cement-based filler to make the repair.

If the whole floor surface is in poor condition, the best cure is to lay a completely new surface screed using a special self-smoothing compound.

What to do

The traditional way of surfacing a concrete floor is to lay a thin screed of fine concrete about 50mm (2in) thick over the floor slab. This has one obvious drawback as far as rescreeding an existing floor is concerned: the rise in floor level will mean repositioning skirting boards, shortening doors and so on.

The solution is to lay an ultra-thin screed using a material known as self-smoothing or self-levelling compound. This is a water-based material which you mix up in a bucket, pour onto the floor surface and trowel out to a thickness of between 3–6mm ($\frac{1}{8}$–$\frac{1}{4}$in). It flows out and finds its own level as you pour it, so can also cope with minor inaccuracies in the floor level. It sets hard enough to walk on in about two hours, and is ready to receive new floor-coverings the next day.

To ensure that the compound bonds well to the old concrete, you must brush off any dust, scrub it with sugar soap and then seal it with an overall coat of diluted PVA building adhesive.

What you need:
For basic repairs:
- concrete sealer
- old paintbrush
- cement-based filler
- cold chisel and club hammer
- filling knife or pointing trowel

To lay a new screed:
- self-smoothing compound
- sugar soap
- PVA building adhesive
- bucket
- scrubbing brush
- old paintbrush
- mixing stick
- plasterer's trowel

CHECK
- that the floor is free from damp before laying the new screed. If any is present, brush on two coats of a bitumen latex waterproofing emulsion first. Check that what you use to damp-proof the floor is compatible with the self-smoothing compound.

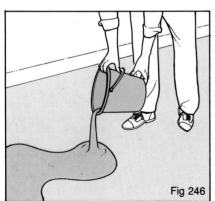

Fig 244 Sweep the floor surface, then scrub it with sugar soap and apply a coat of diluted PVA building adhesive.

Fig 245 Mix up the self-smoothing compound in a bucket, following the maker's instructions.

Fig 246 Pour out the compound onto the floor surface an area at a time, working back towards the room door.

Fig 247 Trowel the surface smooth after pouring each area, and leave to harden.

Patching Damaged Plaster

Plaster is the ideal material in many ways for providing a hard, smooth surface to interior walls, as a basis for decoration using paint or wallcoverings. It has been around in one form or another for centuries, and today's two-coat gypsum plaster systems are direct descendants from the three-coat cement, sand and lime-based plaster favoured in Victorian and Edwardian homes.

Plaster rarely fails through any faults of its own; its problems are usually due to other factors such as defects in the surface to which it is applied or defective skills in selecting and applying the material. It is, however, liable to physical damage which can result in surface cracks and holes, and can be severely weakened by damp – either penetrating through the wall surface or as a result of plumbing leaks.

Even if you never acquire the level of skill needed to plaster a whole wall, you can still tackle small-scale repairs. It is especially easy to tackle patching work, since the surrounding sound plaster still on the wall surface gives you a level to work to.

What to do

If an area of solid wall plaster is badly cracked or damaged, or it sounds hollow when tapped, the best solution is to cut away the damaged or loose material with a sharp brick bolster and club hammer until you reach solid well-bonded material. Undercut the edge of the patch slightly to improve the key when it is filled.

If the existing plasterwork is around 19mm (¾in) thick, it was probably three-coat work built up using cement/sand/lime plaster. Thinner coats around 13mm (½in) thick are likely to be more modern two-coat work using gypsum plasters.

You can repair both types using ready-mixed tub plasters, which have the twin advantages of being available in relatively small quantities, and of drying slowly enough to give the amateur plasterer plenty of time to get a smooth finish. However, they are expensive. The alternative is to patch three-coat work with a cement/sand undercoat and a Sirapite finish coat, and to repair two-coat work with ordinary lightweight gypsum plaster.

What you need:
- brick bolster and club hammer
- stiff brush
- water brush or garden spray gun
- patching plaster
- hawk
- plasterer's trowel
- ruling batten

CHECK
- that the wall is not damp before re-plastering damaged areas. If it is, put the fault right first to prevent a recurrence of the initial failure.
- that the plaster is sound right across the wall if you detect localized patches that sound hollow. It is possible that the whole wall is defective, and wholesale stripping and replastering may be preferable to filling lots of patches.

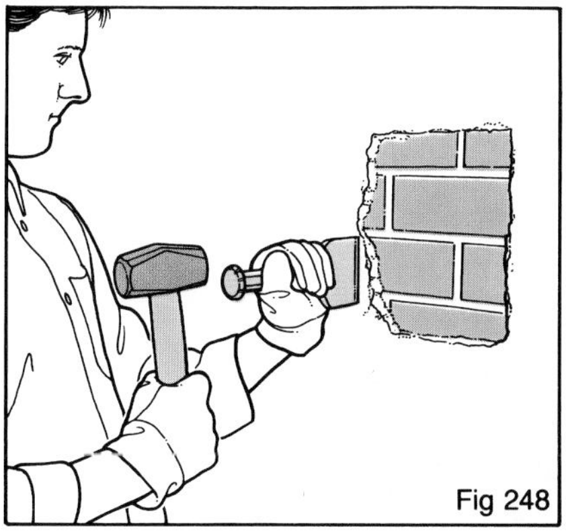

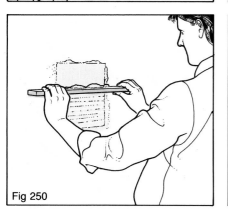

Fig 248

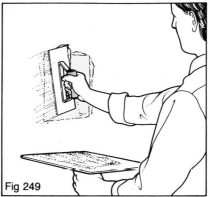

Fig 249

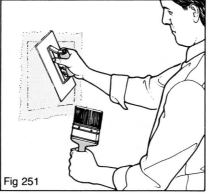

Fig 250

Fig 251

Fig 248 Use a sharp brick bolster and a club hammer to chop away plaster that is damaged or sounds hollow. Cut back until you reach sound material.

Fig 249 With ready-mixed DIY plasters, fill the patch in one go using a plasterer's trowel to press the plaster into the recess. Overfill it slightly.

Fig 250 Use a length of timber batten as a rule to remove excess plaster and leave the patch level with its surroundings.

Fig 251 Finally, polish the repair by flicking water onto the surface and working over it with a steel float.

Repairing Lath-and-Plaster

Until the spread of plasterboard as a building material after the Second World War, almost every home had ceilings formed by applying plaster to the underside of closely-spaced rough timber strips nailed to the ceiling joists. These strips, called laths, were about 38mm (1½in) wide and around 6mm (¼in) thick, and were nailed to the joists with narrow gaps between them. As the plaster was applied, some squeezed up between the laths to form a key which held the plaster in place. The two common faults in such ceilings are failure of the plaster key and physical damage to the ceiling caused by carelessly-handled furniture or, in lofts, carelessly-placed feet.

Lath-and-plaster is also found on the walls of some older properties, especially where the wall masonry is uneven or of poor quality. Here the laths are fixed to timber battens mounted on the wall surface, and the plaster is applied over them as for ceilings. Such walls are obviously also prone to physical damage, and can also suffer from dry rot due to lack of ventilation of the supporting timbers.

What to do

If the key starts to fail in a lath-and-plaster ceiling, tell-tale sagging will be visible and it may be possible to push areas of the plaster upwards against the laths. So long as the ceiling is not cracked, you may succeed in repairing the failed key by pushing the sagging area up (using a board to spread the load and a prop to hold it in place). Then clear away the broken key plaster from the top surface of the ceiling and pour on some quick-setting plaster to form a new key.

If the plaster has cracked badly or has fallen away completely but the laths are still sound, remove all the loose material and replaster the area, forcing the plaster up between the laths to get a sound key. Recess the first coat by about 3mm (⅛in), then apply a finish coat to complete the repair, ruling it off level with its surrounding plaster and then polishing it smooth with the trowel.

Where the laths are damaged or broken, cut back through the ceiling to adjacent joists and fit a plasterboard patch.

What you need:
- quick-setting plaster for forming new key
- patching plaster
- mixing bucket
- hawk
- plasterer's trowel
- batten for rule
- plasterboard patch
- galvanized nails
- padsaw and knife
- hammer

CHECK
- that the key is sound on other parts of the ceiling if one area starts sagging. If it is not, consider pulling down the whole ceiling and replacing it with plasterboard – a better long-term solution to endless repeat patching as other areas fail.

Fig 252

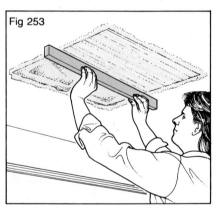

Fig 253

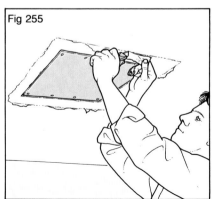

Fig 254

Fig 255

Fig 252   Where the key has failed, pull away the sagging plaster and cut back neatly to well-keyed plaster using a sharp knife. Check that the laths are securely fixed and not split.

Fig 253   Fill the patch in two stages, forcing the undercoat up hard against the laths to form a good key. Then add a finish coat and rule it off level.

Fig 254   If the laths themselves are damaged, cut them back with a padsaw and sharp knife to the centre line of adjacent joists, and remove all broken laths to leave a square or rectangular hole.

Fig 255   Nail up a square of plasterboard to fill the hole, and then plaster over it (*see* page 83 for more details).

Repairing Plasterboard

Plasterboard is one of the most useful building materials to have been introduced this century, enabling ceilings and partition walls to be created in a fraction of the time taken before it arrived on the scene. Its high strength and comparatively light weight plus its ease of fixing have left it with virtually no competition.

However, it does have two weaknesses. It is easily damaged by impact – carelessly-moved furniture, for example, or children driving wheeled toys without due care and attention. It also tends to fall apart when it gets wet, as anyone who has had a plumbing disaster will testify; it will hold the water back for a while, perhaps allowing a trickle through at the joints between the boards, and then fails catastrophically! It is a good job it is easy to replace if disaster strikes.

The one thing that plasterboard needs is proper support all round the board edges, and so when carrying out repair work it is essential to ensure that adequate fixing grounds are provided for even comparatively small sections of board.

What to do

If a partition wall or ceiling suffers impact damage that bursts the board, start by locating the adjacent supporting timbers – the vertical studs in a wall or the joists above a ceiling. Mark these on the board surface. Then draw two lines at right angles to form a neat square or rectangle containing the damage. Use a padsaw to cut along the lines between the joists, and a sharp handyman's knife to cut through the board over the supporting timbers. Remove the panel.

Next, fix 50mm (2in) sq timber noggings between the joists, centred behind the edges of the cut-out so the patching panel can be nailed to them. You can skew-nail these timbers into place, but it is better to use screws driven through angled pilot holes to get a positive fixing.

Measure the size of the panel required, and cut it with a fine-toothed saw. Secure it in place with galvanized nails driven into the supporting timbers and the noggings, then apply a skim coat of finish plaster over the repair.

What you need:
- straight-edge
- pencil
- padsaw
- handyman's knife
- timber noggings
- drill and twist drill
- woodscrews
- screwdriver
- tape-measure
- plasterboard offcut
- fine-toothed saw
- galvanized nails
- hammer
- patching plaster
- bucket and stick
- hawk
- plasterer's trowel

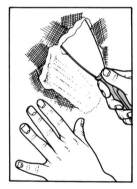

Fig 260   Use fine mesh repair tape to patch small holes, then paint over it.

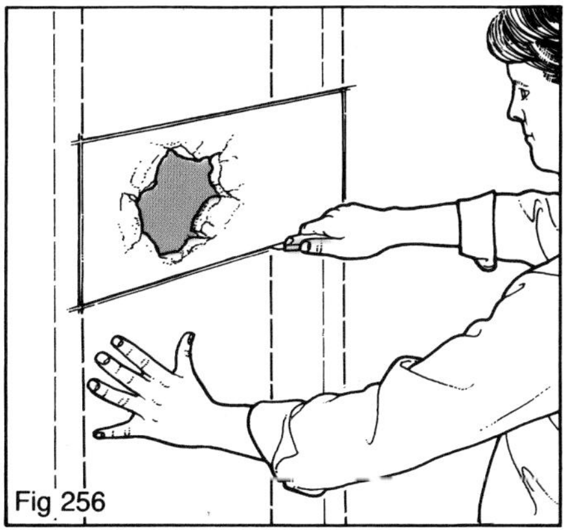

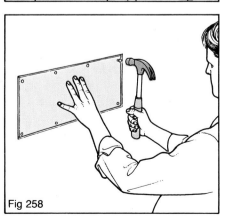
Fig 256

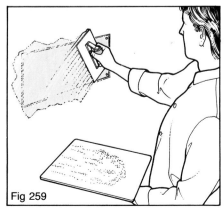
Fig 257

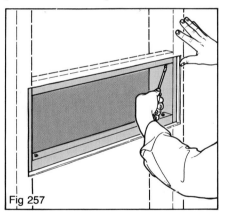
Fig 258

Fig 259

Fig 256   Locate the supporting timbers, then mark out a neat square or rectangle containing the damaged area. Cut it out with a padsaw and handyman's knife.

Fig 257   Screw two supporting noggings into position.

Fig 258   Nail the patch to the noggings and the supporting timbers.

Fig 259   Apply a thin skim coat of finish plaster over the repair.

Restoring Plaster Mouldings

Many Victorian and Edwardian homes have ornate plaster cornices and ceiling centres in their main 'public' rooms – the hall, the landing and in sitting and dining rooms. Some had such decorations throughout the house, although they were often plainer in lesser rooms. These decorations were generally of fibrous plaster; plainer types were often formed *in situ* by skilled plasterers, while others were cast in lengths in moulds and then secured in position through wooden fixing grounds.

Many of these decorations are works of art, with an incredible wealth of detailing, and even plainer types have a smoothness and mass that modern imitations cannot approach. Unfortunately, much of this detail has been obscured by overzealous painting as the years have gone by, to the point where sharp detail is reduced to nothing more than soft curves. Worse still, many have been damaged by the construction of partition walls to subdivide the large rooms typical of the period, or just by carelessness and neglect. They are details that are well worth saving.

What to do

Start by assessing the scale of the problem. If the cornice is intact but merely clogged with paint, it is generally possible to remove the paint so as to expose the original detail, although this can be a long and fiddly job. If it is damaged you may be able to copy small sections of plain types, or to replace them with modern copies; fortunately several firms specialize in producing plaster cornices in the most popular traditional styles, so finding a match may not be as difficult as you think. It is even worth contacting architectural salvage firms in your area; it is possible that examples of similar work may have survived in other houses in your area which have been stripped of their fittings before being demolished.

If you decide to tackle the restoration of an overpainted cornice or ceiling centre, your biggest problem lies in identifying the paints used on it. Recent paint will probably be emulsion paint, but earlier layers could be anything from eggshell to distemper.

What you need:
For restoration work:
- paint stripper
- garden spray gun
- set of improvised picks and scrapers
- soft brush
- plaster of Paris
- filling knife
- endless patience!

For replacement work:
- salvaged cornice sections and ceiling centres *or* modern copies
- galvanized screws
- galvanized washers
- wallplugs
- drill and drill bits
- screwdriver

TIP
Working for long sessions from steps is very tiring. It is better to hire enough platform tower components to enable you to build a mobile platform on which you can sit or even lie while you work.

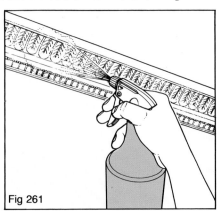

Fig 261

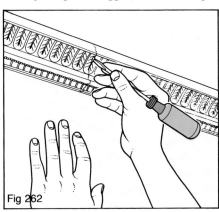

Fig 262

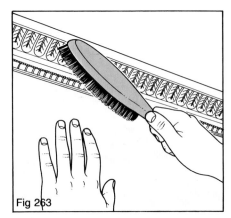

Fig 263

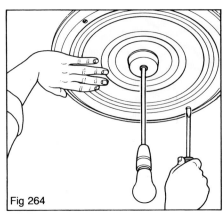

Fig 264

Fig 261   Start by spraying water onto the encrusted paint to see it it softens. If it does not you will have to experiment with chemical paint strippers.

Fig 262   Use improvised picks and scrapers to remove the paint from recesses in the plaster.

Fig 263   Brush loosened particles of material away as you work.

Fig 264   Secure replacement ceiling centres and plaster cornices in place with galvanized screws and washers to ensure a firm fixing.

EXPENSIVE JOBS

However competent a do-it-yourselfer you are, there will be some jobs that are either too big for you to tackle or that require professional skills and equipment. Examples include such major tasks as building a home extension, laying a large driveway (not necessarily beyond your skill, but likely to take too long if tackled on a DIY basis), building high boundary or earth-retaining walls, or rendering or plastering large areas. In all these cases you may prefer to call in an outside contractor to do the work for you. This is a move that many people regard with some trepidation, since finding good, reliable tradesmen can be difficult and choosing the wrong firm could result not only in a bodged job but in considerable financial loss as well.

Here are some general guidelines to help you minimize the risk of picking a cowboy, followed by a look at what is involved in employing professionals.

Finding a Contractor

Once you have decided you need to call in a professional, your first step is to get people with the skills you require to visit the site and give you a firm quotation for the job.

Personal recommendation is by far the best and safest way of finding someone suitable. If a firm has already carried out work for friends, relatives or neighbours you will be able to get a first-hand account of its performance and even to check up on the standard of workmanship.

If this does not work, your next step is to take a walk or a drive round your area, looking for signs of someone carrying out the sort of work you want done. Many firms now put up a sign outside the site they are working on (or park their vans close by), and will not mind if you approach them. You can do this directly, or you can telephone the number on the sign/van. You can also approach householders directly if it is obvious that they have recently completed work similar to what you need doing. Most people are only too happy to show off a job well done and to put you in touch with the contractor who carried out the work.

Next, try your Yellow Pages or Thomson Local telephone directory. Both list local contractors by trades, and many of the display advertisements not only give more details of the sort of work undertaken, but may also reveal whether the firm is a member of a relevant trade association. With this method, it is well worth asking the firm about other jobs it has done locally. Any company worth its salt will be pleased to put you in touch with satisfied customers for whom they have worked.

Your last method of contact with local contractors is via the various professional and trade associations to which many reputable companies and individuals belong. These associations will give you the names and addresses of their members working in your area, and some offer other back-up services such as guarantees and arbitration schemes which may be worth knowing about. Membership of such a body is generally a good sign (many require evidence of several years' trading and satisfactory accounts before granting membership), but it is wise to check that firms actually do have genuine membership of the body concerned – some firms simply 'borrow' logos and claim membership to enhance their image. For more details about individual trade associations, see page 94.

Fig 265 (*above*) Many do-it-yourselfers would prefer to leave larger-scale building jobs to a professional.

Quotations

Getting Quotations

Once you have contacted someone who sounds interested in carrying out whatever work you want done, your next job is to explain clearly what the job involves and to find out as precisely as possible what it is going to cost you, when work can start and how long it will take to complete. For major projects such as building an extension or remodelling your garden it is essential not to rely on verbal agreements, but to ensure that everything is in writing. This can save a lot of arguments, and will also help a court to sort a dispute out if things go seriously wrong. First, make sure you understand the meaning of the following terms, so you know what you are asking for and what the contractor intends you to get.

• *Estimates* are just that – an educated guess as to the rough cost of the job. They are not legally binding.
• *Quotations* are firm offers to carry out a specified job for an agreed price.
• *Tenders* are also offers to carry out specified work for a named price, but are understood to involve an element of competition with other contractors.

Most contractors will want to make a site visit to assess the scale of the job involved before even giving an estimate. Explain in as much detail as possible what you require, and tell him you require a firm quotation for the work, plus details of when he will start and how long the job will take.

Ask at this point whether he or the firm is registered for VAT, and if so whether VAT is payable on the work you are having done. Generally speaking, you do not have to pay VAT on work involving the construction, alteration or demolition of a building, but VAT *is* payable on repair works and maintenance. If he is not a registered VAT trader (with a registration number printed on his notepaper), he cannot charge you VAT on work he does for you.

Always get at least two quotations for the job, and more if you can. This allows you to compare terms as well as prices before making your choice. Impossibly high quotes rarely mean you will get top-quality work; they are the contractor's way of saying he does not want the job, but will do it if you are prepared to pay a silly price.

Assessing Quotations

Once you have received the quotations, study them carefully. The amount of detail given will vary from firm to firm, but the ideal quotation should cover the following points:

• a description of the work to be carried out, preferably setting out all the stages involved (known as a schedule).
• details of materials or fittings to be used, and who will supply them.
• who will be responsible for obtaining any official permission needed.
• when the work will start.
• when the work will be completed.
• who will be responsible for insuring the work and materials on site (professional contractors have both employer's liability and public liability insurance).
• whether subcontractors will be used, and if so, for which parts of the job.
• how variations to methods, timing or costs will be agreed.
• the total cost of the work.
• when payment will be required.

These details form part of the contract between you and whoever you decide to employ, so it is important that they are discussed and dealt with now, to prevent arguments later. Some contractors may include them on a standard form of contract sent with the quotation, or may print their terms and conditions on the reverse of their quotation. In either case, read them carefully; now is the time to discuss any variations or delete any clauses you do not want to apply.

Once you have received quotations from the various firms you approached, it is up to you to decide which one to accept on the basis of price, timing and other factors such as your personal impressions or any recommendations you have received. When you have made your choice, write and accept the quotation.

You now have a contract between yourself and the contractor. In most circumstances, there is no reason to suppose that anything will go wrong, but if it does, tackle it immediately so things can be put right. Mention problems verbally first of all, and if this does not resolve matters, follow up with a letter outlining the nature of your complaint and requesting specific action to correct it.

CHECK
• that your contract covers the following points (where relevant):

1 A detailed description of the work to be carried out.
2 The projected start and finish dates.
3 What will happen if work overruns the projected date in terms of compensation.
4 That the contractor will be responsible for clearing up the site.
5 That the contractor will comply with the requirements of all relevant rules, regulations, laws and bye-laws.
6 That any variation to the work will be agreed in writing by both parties.
7 That the contractor will arrange third party and employer's liability insurance.
8 That the contractor will put right any damage he causes.
9 That the contractor will put right any defects in his work which occur within a set period after completion.
10 How payment will be made for the work.
11 That an amount will be retained until the end of the period agreed for the correction of defects.

FACTS AND FIGURES

This section is intended as a handy reference guide to the range of building and related products you will need to carry out the various jobs described earlier in the book. It will help you to see at a glance what is available, and in what sizes or quantities, so you can plan your requirements in detail and draw up itemized shopping lists for individual projects.

Bricks

Clay and concrete bricks are available in a huge range of colours and textures, but thankfully all come in one common standard size – 215 × 102.5 × 65mm (8¼ × 4 × 2⅝in). This is the actual overall size of the brick; for the purposes of estimating materials you should use what is called the work size of 225 × 102.5 × 75mm (8⅞ × 4 × 3in) to allow for vertical and horizontal mortar joints 10mm (⅜in) thick. So you will need sixty bricks per square metre of wall 102.5mm thick, and 120 per sq m for a wall 215mm thick.

Concrete bricks are also available in a range of metric sizes. The commonest actual sizes are 190 × 90 × 65mm (7½ × 3½ × 2⅝in), 190 × 90 × 90mm (7½ × 3½ × 3½in) and 290 × 90 × 90mm (11⅜ × 3½ × 3½in), giving work sizes including mortar joints of 200 × 100 × 75mm, 200 × 100 × 100mm and 300 × 100 × 100mm respectively – far easier to work with than the traditional brick size when it comes to estimating quantities. Per square metre of wall you will need sixty-seven, fifty and thirty-four bricks respectively for the three sizes mentioned.

Bricks are sold by number, so in theory you can order the precise number you need for a particular project. However, most suppliers deal in 100s and 1000s, and will give a better price for an order that is a round number. In any case, you should always overestimate quantities by about five per cent to allow for breakages.

Concrete blocks – the squared-off type used in housebuilding, not the ornamental garden walling block – are useful for a range of projects where the finished masonry will be hidden, such as the back skin of an earth-retaining wall. The table (*right*) lists some typical sizes.

Length × height of concrete blocks			
Work size (mm)	Actual size (mm)	Actual thickness (mm)	No per sq m
400 × 200	390 × 190		12.5
450 × 150	440 × 140		15
450 × 200	440 × 190	60, 75, 90,	11
450 × 225	440 × 215	100, 140, 150,	10
450 × 300	440 × 290	190, 200 & 215	7.5
600 × 150	590 × 140		11
600 × 200	590 × 190		8.3
600 × 225	590 × 215		7.5

Walling Blocks

Walling blocks – the reconstituted stone type, man-made to imitate natural stone – come in a bewildering range of sizes. It seems that each manufacturer has his own favourite, although it is fairly common to find blocks in most ranges measuring 230 or 300mm (9 or 12in) long, 100 or 150mm (4 or 6in) wide and 65mm (2⅝in) high – the same height as a standard brick. Others in the range are usually multiples of this, which allows you to lay the blocks in a wide range of different decorative bonds.

Some manufacturers also offer larger blocks which have their outer face moulded so they resemble a number of smaller, randomly-shaped blocks. The moulded joints are deeply recessed, and each block may have projecting units at each end to allow a stretcher-style bond to be achieved between neighbouring blocks.

The best way of estimating quantities of garden walling blocks for any particular project is to select the block you want to work with and then to use its actual size as a guide – look at the table (right), which covers some popular sizes. To work out how many blocks you will need per square metre of single-thickness wall for blocks not included there, add 10mm (½in) to the actual block length and height (in mm), multiply the two figures together and divide the result into 1,000,000 (the number of sq mm in a sq m). So for a 440 × 65mm block the sum is: 450 × 75 = 33,750sq mm, then 1,000,000 divided by 33,750 = 29.63, which means you need thirty blocks per sq m. It helps to have a calculator!

Screen walling blocks are widely available in just one standard size – an actual size of 290mm (11⅜in) square and 90mm (3½in) thick, giving a work size of 300mm (11¾in) square. You will need eleven blocks per square metre of wall. Coping stones to finish off the top of the wall usually come in 610mm (2ft) lengths – long enough to bridge two of the walling blocks and three 10mm (⅜in) mortar joints.

The special hollow pier blocks that all manufacturers make as part of their screen walling block range are generally about 200mm (8in) square and 190mm (7½in) tall. Allowing for a 10mm (½in) mortar joint, this means that three pier blocks build up to the same height – 600mm or just under 2ft – as two walling blocks.

Length × height of walling blocks		
Work size (mm)	Actual size (mm)	No per sq m
230 × 75	220 × 65	58
300 × 75	290 × 65	45
300 × 145	290 × 135	23
300 × 225	290 × 215	15
300 × 300	290 × 290	11
450 × 75	440 × 65	30
450 × 112	440 × 102	20
450 × 150	440 × 140	15

Fig 267 (*above*) Garden walling blocks come in a wide range of shapes, sizes, textures and colours, and generally have matching coping stones and pier caps.

Fig 268 (*below*) Screen walling blocks come in a small range of designs, and are usually 290mm (just under 12in) square.

Paving Slabs and Blocks

Paving Slabs and Blocks

Reconstituted stone *paving slabs* are usually squares or rectangles, and come in a range of sizes based on a 225mm or 300mm (9 or 12in) module. The commonest sizes are listed in the table below, along with the numbers needed to cover an area of 10sq m (about 12sq yds).

Slab size (mm/in)	No per 10sq m
225 × 225mm (9 × 9in)	200
300 × 300mm (12 × 12in)	110
450 × 225mm (18 × 9in)	100
450 × 300mm (18 × 12in)	75
450 × 450mm (18 × 18in)	50
600 × 300mm (24 × 12in)	55
600 × 450mm (24 × 18in)	37
600 × 600mm (24 × 24in)	27
675 × 450mm (27 × 18in)	33

Use these figures as a rough guide to estimating the scale (and cost) of your projects. In practice it is easier to design paved areas if possible so they are a whole number of slabs wide and long, to minimize the need for cutting slabs. Then you can count how many rows there are and how many slabs are needed in each row, and multiply the two figures. If you are laying mixed slabs of different sizes, draw a scale plan on squared paper and count how many of each size slab are needed.

Some manufacturers also offer hexagonal and circular slabs. The former are usually 400mm (16in) wide, measured between two opposite parallel sides, and come with matching straight-sided half slabs to allow you to pave square or rectangular areas with them. You will need fifty-five 400mm hexagons to pave an area of 10sq m. Circular slabs come in several diameters from 300mm (12in) upwards, and are intended to be laid as individual stepping stones.

Cheaper cast concrete slabs are usually 50mm (2in) thick, while the more expensive hydraulically-pressed types are generally 40mm (1⅝in) thick. They are surprisingly heavy – a 450mm sq slab weighs around 16kg (36lb, or 2½ stone) – so lift them with care. Do not load the back of your car with more than a few of them either; you will seriously compromise the car's steering, and may ruin the rear suspension into the bargain. Have your order delivered by your supplier.

Paving blocks are generally rectangular in profile, although some are made in interlocking shapes to give a less regular look to the surface of the paved area. The standard block size is 200 × 100mm (8 × 4in) so estimating coverage is easy; you need fifty per sq m of surface area. Most are 65mm (2½in) thick; some light-duty blocks are only 50 or 60mm (2 or 2⅜in) thick.

Fig 269 (*above*) Paving slabs are generally based on a 225 or 300mm (9 or 12in) module.

Fig 270 (*below*) Embossed finishes and unusual shapes add interest to paved areas.

Mortar and Concrete

The basic ingredients of mortar and concrete are cement, aggregate of one kind or another, and various additives that improve the performance or ease of handling of the mix. Estimating the quantities you need for individual jobs can be very difficult for the amateur builder, particularly since these materials have to be ordered in 'trade' quantities for all but the smallest jobs, and waste costs money. The information given here will, it is hoped, make it easier to estimate your requirements for individual projects.

Cement is generally sold only in 50kg (1cwt) bags, although many DIY superstores retail it in smaller and more manageable 25 or 40kg packs. All the mortar and concrete mix formulae on page 91 are based on the use of standard 50kg bags.

Sand, both soft (building or bricklaying) and sharp (concreting) varieties, is sold either in bags or by volume. Bags – either 40 or 50kg – are convenient for small jobs, but work out extremely expensive for large projects such as laying patios or concrete slabs. For these it is best to order by volume from a builders' merchant. The smallest quantity most will deliver is half a cubic metre, or about three-quarters of a ton. Beware the firm which still speaks of cubic yards and tells you they are the same as cubic metres. They're not: a cubic metre is some 30 per cent bigger, and this discrepancy could throw out your calculations quite seriously.

Aggregate is also sold bagged or by volume, and is graded according to the size of the average particle – fine if it will pass through a 5mm (¼in) sieve, and coarse otherwise. Coarse aggregates for concreting usually have a maximum stone diameter of 20mm (¾in), although you can get 10mm (⅜in) aggregate for use in fine concrete. A cubic metre weighs nearly two tons, and is a surprisingly large heap on the drive!

Lime for use as a mortar ingredient usually comes in 25kg (55lb) bags. Other additives such as plasticisers and frostproofers are generally sold in 1, 5 and 25 litre containers, while pigments are packed by weight – commonly in 625 gram and 1.25kg (just under 3lb) sizes.

Fig 271 (*left*) Mortar and concrete ingredients – pigments, sand and aggregate, cement and chemical additives.

Mortar and Concrete

Mortar and Concrete Formulae

For mortar, use Table 1 to select the mix you need according to the job you are tackling, then use the formula for the mix as detailed under *mix types*. For concrete, use Table 2 to do the same. The figures in column 3 are amounts needed to make 1cu m of concrete. Note that all mixes should be proportioned by volume, using a bucket.

1: MORTAR MIXES

Use	Exposure	Mix/sand type
Bricklaying and blocklaying using dense aggregate blocks	Moderate Severe	Mix B soft Mix A soft
Blocklaying using aerated or light aggregate blocks	Moderate	Mix B soft
Pointing bricks and dense aggregate blocks	Moderate Severe	Mix A soft Mix C soft
Pointing aerated and light aggregate blocks	Moderate	Mix B soft
Rendering on brick or dense aggregate blocks	Moderate	Mix A sharp
Rendering on aerated or light aggregate blocks	Moderate	Mix B sharp
Thin screeds (20–40mm) on dense concrete	Internal	Mix C sharp
Thick screeds (50–75mm) on dense concrete	Internal	Mix D sharp

2: CONCRETE MIXES

Use	Proportion by volume		Amount per cu m	Yield per 50kg bag of cement
General purpose Most uses except foundations and exposed paving	Cement Sharp sand 20mm aggregate OR All-in ballast	1 2 3 4	6.4 bags 680kg/0.45cu m 1,175kg/0.67cu m 1,855kg/0.98cu m	0.15cu m
Foundations Strips, slabs and bases for precast paving	Cement Sharp sand 20mm aggregate OR All-in ballast	1 2½ 3½ 5	5.6 bags 720kg/0.5cu m 1,165kg/0.67cu m 1,885kg/1cu m	0.18cu m
Paving All exposed surfaces, all driveways	Cement Sharp sand 20mm aggregate OR All-in ballast	1 1½ 2½ 3½	8 bags 600kg/0.42cu m 1,200kg/0.7cu m 1,800kg/0.95cu m	0.12cu m

Ready-mixed concrete
When ordering ready-mixed concrete, be sure to specify:

- the volume of material you need
- what it will be used for (i.e. foundations, a driveway etc)
- when you want it
- how you will handle the delivery
- whether there is easy access to the delivery site. Most trucks are about 8m long and 2.5m wide, and can discharge their contents via a shute within about 3m (10ft) of the back of the truck.

Plaster and Plasterboard

Plaster, unlike cement, comes in a somewhat confusing range of types. They can be divided into two grades; one is used to form relatively thick base coats or undercoats, the other for comparatively thin finish coats. The vast majority of the plastering work you are likely to carry out in the home will involve using gypsum-based plasters such as Carlite, Thistle and Sirapite. Cement-based plasters are used mainly in renovation work where damp or other conditions preclude the use of gypsum plasters.

It is important to choose the correct undercoat plaster if you want to achieve good results, and which one you select depends on the porosity of the background to which it will be applied. On surfaces with what is known as normal suction – most bricks and blocks, for example – you should use Carlite Browning plaster; low-suction masonry needs Carlite Bonding plaster. Use special Metal Lathing plaster for undercoats over metal lathing and expanded metal mesh.

For the finish coat, use Carlite Finish plaster over Carlite undercoats, and Thistle Board Finish on plasterboard. If you are finishing a cement-based undercoat, use either Thistle finish or Sirapite.

Plaster is widely sold in 50kg (1cwt) bags, and DIY superstores and other retail outlets also stock smaller sizes which are useful for repair work. Note that plaster does not keep well, and should not be stored for more than two months. Look out for the manufacturer's date stamp on the sacks when buying.

Coverage obviously depends on coat thickness, but as a guide 50kg of Browning plaster will cover about 7.5sq m (9sq yd), while Bonding plaster will cover up to

8 sq m (9½sq yd). Carlite Finish covers up to 25sq m (30sq yd), while Thistle Board Finish, applied as a 5mm thick coat, will cover about 8sq m.

One-coat plasters, designed to be applied as a combined undercoat and finish coat, will cover about 4.5sq m (5½sq yd) at a thickness of 12mm (½in).

Ready-mixed tub plasters do not go very far – a 6-litre tub of undercoat will cover less than 1sq m (11sq ft), while the same quantity of the finish coat will cover about 3sq m (3¾sq yd) – so they are an extremely expensive way of plastering more than a small area of wall.

Fig 272 (*above*) Plasterboard is available plain, or with a vapour barrier or a layer of insulation bonded to one surface.

Fig 273 (*far left*) Gypsum plasters such as Carlite, Thistle and Sirapite are sold in bags ready for mixing with water.

Fig 274 (*left*) Ready-mixed undercoat and finish plasters come in tubs.

Plaster and Plasterboard

Plasterboard

Plasterboard for surfacing ceilings and the timber frames of stud partition walls comes in several types and a wide range of sheet sizes. The sheets consist of a core of lightweight plaster, covered on both sides by a strong paper liner. Surfaces with a grey finish are intended to be plastered over, while those with an ivory finish can be decorated directly with paint or wallpaper. Ivory-faced boards may have a tapered edge so that joints can be taped and filled flush.

Standard plasterboard or wallboard, with one grey and one ivory face, is by far the most widely available type. The commonest sheet size is 2.4 × 1.2m (8 × 4ft), but other sizes are made and can be ordered if not in stock in widths of 600 and 900mm (2 and 3ft) and in various lengths from 1.8m (6ft) up to 3m (10ft). The longer lengths are useful if you are working in rooms with higher-than-average ceilings. Wallboard is available in 9.5 and 12.7mm (⅜ and ½in) thicknesses.

Baseboard is used mainly for lining ceilings which will be skim-plastered. It is available only as a square-edged board and has grey paper on both faces. It comes in lengths of 1.2, 1.219, 1.35 and 1.372m (3ft 11¼in, 4ft, 4ft 5⅛in and 4ft 6in respectively), and in a standard width of 915mm (3ft).

Both baseboard and wallboard are also available as Duplex boards with a metallized polyester film backing which acts as a vapour barrier; this prevents moisture vapour being transmitted through the board and causing condensation within the structure to which the board is fixed.

Moisture-resistant board is widely used in timber-framed buildings, and as a sheathing material behind external wall tiling, weatherboarding and other claddings. It can also be used as a lining in areas such as porches where moisture vapour levels may be high, and on walls which will subsequently be decorated with ceramic tiles – in shower cubicles, for example. It comes in two sizes, both 9.5mm thick: 1.8m × 600mm (6 x 2ft) and 2.4 × 1.2m (8 × 4ft). The larger sheets are also available in the 12.7mm (¾in) thickness.

Thermal board is insulating plasterboard with a layer of polystyrene or urethane foam bonded to the rear face, and usually incorporates a vapour barrier too.

Estimating from Plans

On all but the smallest building projects, drawing up working plans is an essential first step. Not only will these help you visualize what the end result will look like and make it easier to plan the stages by which work will proceed; they are also an invaluable aid to estimating materials.

Taking as an example the garden plan featured on pages 32 and 33 and reproduced below in its finished form, you can see how helpful such a drawing can be. Here one square of the grid has been taken to represent one 18-inch square paving slab, so immediately revealing the need for 180 slabs. As for the dwarf brick walling, the plan reveals a total run of approximately 27m (90ft) of wall three courses high, with a soldier course on top. So the walls will need 120 bricks per course laid in stretcher bond – a total of 360 bricks plus a further 360 for the soldier course, making a total of 720. Even estimating quantities of sand for bedding the slabs is easy – 36sq m at a depth of 50mm (2in) means just under 2cu m of sand.

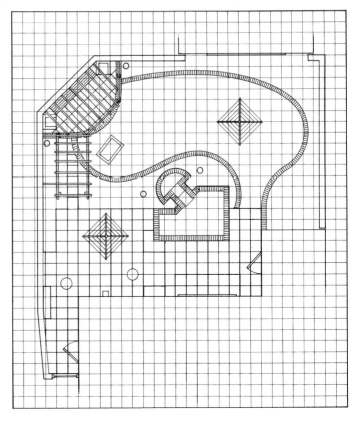

Fig 275   Use scale drawings of your proposals as a guide to estimating quantities.

Glossary

Additives Chemicals added to mortar and concrete mixes to improve their performance. The commonest include plasticisers, waterproofers and frost-proofers.

Aggregate Sand or gravel added to cement to make mortar or concrete respectively. The term is normally used for all-in aggregate, a mixture of sand and gravel mixed with cement to make concrete.

Basketweave bond A pattern used when laying paving bricks and block pavers, consisting of pairs of bricks which are laid at right angles to each other.

Bat A brick cut down in length – a half-bat or a threequarter-bat – for use in maintaining the bond in a wall.

Bond The arrangement of bricks in a wall, designed to increase its strength as well as to enhance its appearance.

Builder's square A rough wooden triangle made up from sawn timber with sides in the ratio 3:4:5 and used to check the squareness of brickwork and concrete.

Cement The adhesive from which mortar and concrete are made. Portland cement is the commonest type. It is usually sold in 50kg (1cwt) bags, although smaller sizes are also available.

Concrete A mixture of cement, sand and gravel (all-in aggregate), used to cast foundations, drives and base slabs. It can be mixed by hand from dry ingredients, or ordered ready-mixed if large quantities are required.

Coping stones Flat or ridged stones used to weatherproof the top of brick and screen block walls. A coping can also be formed by a course of bricks laid on edge – known as a soldier course.

Damp-proof course Layer of waterproof material built into walls rising off the ground to prevent damp from rising in the masonry. In older homes, slate was often used, but modern homes have DPCs of bituminized felt or plastic. Faulty DPCs can be cured by injecting special chemicals into the brickwork.

English bond A bricklaying bond consisting of alternate courses of stretchers and headers.

English garden wall bond A bricklaying bond used mainly for garden wall construction. It consists of up to five courses of stretchers interspersed with a single course of headers.

Flashing Strip of waterproof material used to waterproof the join between a roof and an adjacent vertical surface.

Flaunching The sloping mortar fillet round the base of a chimney pot, sealing it to the stack.

Flemish bond A bricklaying bond in which each course consists of a pair of stretchers followed by a single header.

Formwork Timber used to support the edges of concrete slabs, paths etc. while the concrete is placed. The boards making the formwork are nailed to stout pegs.

Foundations Cast concrete strips or slabs laid to support walls and buildings.

Hardcore Broken brick, concrete etc. used as infill and support beneath concrete foundations.

Herringbone bond A pattern used when laying paving bricks or block pavers. Each brick is laid at 90° to its neighbour, and overlaps it by half the length of the brick.

Manholes Chambers installed on drain runs where branch drains enter the main run or the run changes direction, usually fitted with a rectangular or round metal cover.

Masonry cement Ordinary Portland cement with added plasticiser.

Mastic A non-setting filler used to seal gaps between building components such as frames and masonry.

Mortar A mixture of sand and cement with added plasticiser, used for bricklaying and rendering.

Open bond A bricklaying bond formed by leaving a space between the bricks, which are laid end-on in stretcher fashion.

Pier A thickened section of masonry built at the ends of a wall and at intervals along its length for extra support.

Pointing The finish given to the mortar courses in bricklaying. Several different profiles are used.

Running bond A pattern used when laying paving bricks and block pavers. The bricks are laid end-to-end throughout, imitating stretcher-bond brickwork.

Stack bond A blocklaying bond used chiefly with square screen walling blocks.

Stretcher bond A bricklaying bond consisting of a single skin of bricks 112mm (4½in) thick, with bricks overlapping each other by half their length.

Toothing Process of cutting out bricks in one wall so that those in a wall at right angles to it can be bonded into the original structure.

Weepholes Unpointed gaps left between the bricks in retaining walls to allow drainage to occur through the wall.

Useful addresses

Builders

Building Employers
Confederation,
82 New Cavendish St,
London W1M 8AD
Tel: 071–580 5588

Federation of Master
Builders,
33 John St,
London WC1N 2BB
Tel: 071–242 7583

Damp treatment firms

British Wood Preserving
& Damp Proofing
Association,
6 The Office Village,
4 Romford Road,
Stratford,
London E15 4EA
Tel: 081–519 2588

Information

The Brick Development
Association (BDA),
Woodside House,
Winkfield,
Windsor,
Berkshire SL4 2DX
Tel: 0344 885651

The British Cement
Association,
Wexham Springs,
Wexham,
Slough,
Berkshire SL3 6PL
Tel: 0753 662727

The British Ready-
Mixed Concrete
Association,
1 Bramber Court,
2 Bramber Road,
London W14 9PB
Tel: 071–381 6582

Index

Index